AI on the Frontlines

Cyber Defense and Offensive Strategies for the Digital Age

GEW Intelligence Unit

Global East-West. London

Copyright © 2025 by GEW Intelligence Unit

Essays, Analyses and Reports: A Global East-West Series.

Under the supervision of Dr Hichem Karoui.

All rights reserved.

No portion of this book may be reproduced in any form without written permission from the publisher or author, except as permitted by copyright law.

Contents

1. AI On The Frontlines Cyber Defense And Offensive Strategies For The Digital Age — 1
2. Introducting AI's Role in Modern Cybersecurity — 21
3. Foundations of AI in Cyber Defense — 43
4. Building Intelligent Defense Systems — 63
5. AI-Driven Threat Intelligence and Prediction — 85
6. Automated Incident Response & Recovery — 107
7. Offensive AI: The Cyber Attacker's Arsenal — 131
8. Autonomous Exploitation and Penetration Testing — 151
9. Case Studies: Battles in the Wild — 171
10. Ethics, Law, and Policy in AI-Powered Cyber Warfare — 191
11. The Future Landscape: Towards Autonomous Cyber Conflict — 211

Appendices

1

AI On The Frontlines

Cyber Defense And Offensive Strategies For The Digital Age

The Rise of Cyber Warfare: The Start of a New Era

Every decade since the first computer was invented has brought along major developments in technology. Evolving technology has been strained into the avenues of organised conflicts known as warfare. From the Stone Age where physical lethality ruled, to the modern age where virtual lethality becomes the epitome through virus attacks on physical hardware of machines. Evolving technology has easily made resources available to everyone; however, placing it under tight-security regimes has always had a profound effect on any latent threats. Be it malicious hackers or defending AI, through its algorithmic networks, one cyber threat has always led to a defensive solution by changing its face time and again, covertly penetrating countermeasures. Self-learning algorithms offer greater insights than the best mathematicians in the world, able to go through massive data reservoirs, providing complex covert entry for obstructions. Armed with innumerable nimble drones capable of inflicting selective strikes on critical zones, the use of AI is bound to cause life as we know it to ultimately go down in a cataclysmic event. Newfound threats for old ways of solving bog problems, as the war on resource warfare is bit by bit cutting man out of the equation, simultaneously blurring the lines of virtual and physical warfare.

Fundamental Concepts of AI-Enhanced Defence Strategies

These key principles help protect the digital assets and the critical infrastructure of an institution in modern cyber warfare. AI (Artificial Intelligence) powered proactive and adaptive defences have emerged as crucial in contemporary warfare. It is important to comprehend the numerous levels that constitute AI enhanced defence strategies. To begin with, proactive threat recognition and action prevention constitute a part of AI enhanced defence. AI performed at higher levels can utilise advanced algorithms to study network traffic patterns, discover anomalies, and even predict and avert potential threats before they happen. With such foresight, defenders will always be at least a step ahead of their opponents. AI-based adaptive response initiatives also play a vital role in defence strategies. Automated incident response is made possible with AI technologies, ensuring that responses address the constantly changing landscape of cyber threats in real-time. AI powered defence systems adapt to new threats as they arise, evolving their capabilities in real time. AI allows for ultra-fast decisions to be made as it can correlate and prioritise security events on its own, based on relevance and context, which sharpens the overall agility and responsiveness of cybersecurity operations.

Moreover, the application of AI to defence systems highlights the need for comprehensive situational awareness. With the help of AI analytics, defenders better understand their environments digitally, and as a result, are able to notice signs of compromise and emergent attack behaviours that are more difficult to detect. This enhances the defender's visibility which increases the accuracy and effectiveness of threat detection. Thus, organisations can identify and neutralise threats proactively across their networks.

AI's ability to contextualise and correlate large volumes of disparate information enhances accuracy in detection and attribution of threats, automating investigative tasks, and lowering the rate of invalid alerts. Finally, AI-enhanced defence strategies focus on the integration of human domain knowledge with AI technologies. The synergy between humans and machines in the collaboration human and machine teaming leads to the situation where the analyst's skills are improved with the help of AI, thus allowing him to devote his energies to higher-level strategic work while AI takes care of mundane, repetitive tasks. Such collaboration improves cyber defence operations as far as their adaptability, efficiency, and overall operational resilience are concerned. To conclude, the main points of issues discussed here have been organised around the needs of AI-enhanced defence strategy focusing on proactive threat anticipation and identification, adaptive response protocols, precise post detection responses, multi-source data fusion and information mining, AI and human tackle issues together interactions.

Machine Learning Approaches to Threat Identification

In the realm of cybersecurity, the identification and mitigation of potential threats is a multi-faceted challenge that arises from the modern technological landscape. One important approach that has been adopted is the application of machine learning (AI) techniques to bolster threat detection and response frameworks. Armed with data-driven algorithms, organisations can detect and respond to proactive threats in a timely manner. In this regard, supervised learning algorithms offer one powerful approach of training on labelled datasets to detect patterns as well as deviations that may signify some form of hostile activity within the system. On the contrary, systems that employ unsupervised learning techniques to recognise new threats are increasingly important, especially in the context of identifying and responding to sophisticated emerging attacks, novel without predefined labels. Reinforcement or reactive learning can be used to adapt cyber defence mechanisms relative to persistent and evolving attacks, defences, and countermeasures. As far as machine learning models tailored to cybersecurity are concerned, feature engineering is equally critical because it refers to the extraction and transformation of specific dimensions relevant to threat data improvement. Furthermore, stronger protection against aggressive evolving cyber threats is possible with ensemble learning techniques that fuse several models or classifiers due to enhanced predictive power.

The application of advanced analytical and visualisation techniques alongside algorithms enables comprehensive attack pattern recognition and enhances mitigation efforts. Continuous enhancement of automated threat detection in machine learning-based systems shows the increasing need for research and development through the sophisticated digital threats of today's world.

Current Cybersecurity Issues

In today's world, machine learning and artificial intelligence play a major role in proactive mitigation of security measures. AI is a core element of automation, which enhances anomaly detection in networks, therefore instigating a more rapid and formidable response to intrusions. Providing real-time defence requires constant monitoring along with incoming data stream evaluation. AI-enabled algorithms perform the critical function of detecting pervasive and often subtle trends or indications that signal the possibility of threats. Advanced machine learning methods, particularly deep learning and neural networks, enable milliseconds filtering through extensive datasets in seconds to isolate potentially harmful abnormal changes through deviation flagging.

Anomaly detection is critical in real-time data analysis, allowing systems to detect potential breaches or emerging threats. AI-powered alarms based on established patterns can quickly mark out flagged behaviours; AI anomaly detection systems do quickly warn cybersecurity teams to investigate suspicious activities. AI systems act as first-response alert systems, empowering firms with the capabilities to avert imminent aggressive cyber responses and enabling fortification of their networks and defences. Whether it is unusual network traffic increase, file changes, or anomalous user accesses, AI systems are capable of having the upper hand.

The addition of anomaly detection to existing infrastructures will require merging other technologies heavily integrated with AI, thus requiring the other operational technologies of the framework to be crisis-proofed beforehand. This type of implementation requires a wider scope within the specific context of the organisation, such as data sources and network designs. The alerts provided must also be accurate while reducing the number of false positives. Too many of these erroneous alerts can exhaust users, making alert fatigue a concerning problem in distinguishing critical notifications. This emphasises the need for tweaking algorithms applied for each environment as fine-tuning is crucial towards merging AI anomaly detection with network systems architecture; thus, multidisciplinary merging requires precise multi-framework coordination.

Moreover, there is no bound to internal systems for performing anomaly detection and real-time analysis. Such features also include the cloud environment, IoT ecosystems, and third-party integrations. This requires an all-encompassing strategy for countering threats in varying slices of technology. With the application of AI anomaly detection solutions, companies obtain unmatched information concerning gaps in system defences, which allows them to take preventive actions to reduce cyber risks. In the end, an AI-powered infrastructure where real-time data analysis and anomaly detection collaborate marks an aggressive shift in the perpetual struggle to fortify a digital architecture against constantly emerging threats and defend sensitive data.

Artificial Intelligence in Existing Systems

The implementation of artificial intelligence (AI) technologies is transforming cybersecurity, as AI dramatically enhances the precision of threat analysis and mitigation processes. While AI has made significant advancements in automating processes that require intensive data analysis, it is important to remember that incorporating AI into pre-existing systems requires well-defined automation frameworks.

Cultivating trust in AI's decision-making capabilities among cybersecurity professionals is important, building a working relationship that blends AI proficiency with Human Intelligence (HI). Furthermore, companies must create policies governing the interplay between humans and AI, defining the roles and responsibilities associated with AI actions and decisions within the cybersecurity domain.

Equally important is the ethical dimension concerning AI use, especially where compliance with laws on privacy, jurisdiction, responsibility, and due diligence is concerned. Protecting sensitive data and ensuring visibility in AI processes remains commensurate to the regulatory expectations and ethical standards. Enforcing stricter control parameters and ongoing evaluations can mitigate the risks posed by AI integration, in turn reinforcing trust in the cybersecurity infrastructure.

As long as AI is progressing, the integration should be continual and evolutionary in nature so that newer AI improvements and refinements can be incorporated seamlessly. Vigilant evaluation of AI performance in the context of cybersecurity, along with constant knowledge and skill acquisition, will ensure AI will be put to its best use. Therefore, constant recalibration of strategies due to shifting threats and rapid advancements is crucial to fortifying the cyber defence systems.

In the end, the incorporation of AI into the existing structures of cybersecurity systems is contingent on robust tactics that include a multisector approach—policy, technology, organisation, and dynamics.

Building upon existing frameworks with the aid of Artificial Intelligence can strengthen an organisation's defences and allow for proactive adaptation to the evolving cyber threat environment.

AI and Network Security Protocols

Considering the contemporary cyber warfare narrative, the importance of network security protocols is paramount. These protocols are essential for any business as they provide a line of defence against cyber-attacks as well as protect sensitive information and important systems. The field of network security protocols has changed significantly due to the incorporation of Artificial Intelligence (AI). The application of AI to network security protocols has made them more proficient in the identification, mitigation, and response to proactive cyber threats.

AI in network security protocols supported with technologies such as machine learning does not operate with static predefined rules to examine traffic for any movement patterns which are considered normal across the organisation, identify irregular behaviour, and flag possible breaches of security within the systems in real time. The enhancement of AI protocols helps to secure organisational networks by providing adaptive responses to constant shifting environments. These protocols are now able to respond to attacks in seconds and identify positive and negative activities with high speed.

Beyond this, the protocols make use of AI to provide approximate analyses which help organisations understand the gaps and assume possible chances of cyber-attacks allowing them to take corrective decisions. Transforming the way organisations operate, AI is enabling the prediction purposes and actively searching for and noticing newly emerging systems of cyber-attacks and weaknesses. AI does this on a constant basis and goes on monitoring larger system sets of network data for the breaches that need to be fixed before the attackers initiate an attack, and this aids in strengthening the defences of organisations.

AI technologies have advanced automated decision-making in network security, enabling timely and contextual responses to potential threats. Automated decision-making and threat response is possible via AI technology since security protocols can autonomously gauge potential incidents, prioritising actions and modifying defensive responses in reaction to changing cyber threats in real time.

With that said, it is crucial to point out gaps in AI integration into network security protocols. Maintaining robust AI models integrated into the network security protocols requires constant monitoring, training, and refining due to rapid changes in threat expectations.

Equally important, network security protocols powered by AI require heightened scrutiny and regulation to ensure ethical and transparent utilisation of the technology.

As businesses work on implementing AI technologies into network security protocols, they need to focus on utilising the technology to fortify defences while being mindful of the associated risks and ethical implications. Applying AI technologies mindfully within network security protocols will help businesses develop cyber defence systems that are resilient, adaptive, proactive, and aligned with modern cyber threats.

Case Study: Effective Implementation of Artificial Intelligence in Cyber Defence

The use of Artificial Intelligence in cyber defence was the newly implemented strategy in the case of a large company. This corporation employed the most sophisticated machine learning techniques to protect their digital systems. In this study, we analyse the impact that modern AI technologies had on the company's security systems and how they protected the organisation against advanced attacks. The corporation implemented AI-driven systems for anomaly detection which monitored all network traffic for any suspicious activities. These systems were capable of detecting potential intrusions far more accurately than any previously used systems. The advanced behavioural analytics employed enabled the AI system to quickly isolate and neutralise dangerous activities, reducing the risk of data theft or financial fraud.

Artificial Intelligence systems have been integrated into workflows which respond to the incidents; consequently automating a major part of the work of detecting and dealing with security breaches. The system was able to provide real-time correlation analysis and thus study the data to discover relationships among events which are characteristic of system abuse, thus aiding the security personnel to take timely necessary action. The company was able to very effectively combine artificial intelligence and manpower in the resolution of cases which allowed for continued business operations undeterred from the increasing volume of attacks.

Moreover, AI's implementation in cyber defence was successful with the integrated threat intelligence system which used sophisticated algorithms to identify and forecast potential emerging threats. The system analysed multiple datasets to detect patterns and provided critical insights to safeguard assets, all while identifying possible indicators of compromise. Such measures enabled proactive actions to be taken well in advance. Such strategies enhanced the organisation's overall cyber resilience while diminishing exposure to advanced persistent threats, resulting in an evolving capacity to endure cyber assaults and countering cyber threats with no positive and false positive responses.

Equally, AI tools have proven to be underwriting and compliance ratios to primary industry benchmarks and relevance areas. Cyber policy automated enforcement capability resulted in improved policy enforcement and effective cyber governance, while changepoint detection and adaptive modelling provided ongoing defence mechanisms for ever-changing attack surfaces. Therefore, the corporation not only enhanced evasive capabilities, but also bolstered stakeholder assurance by demonstrating proactive commitment towards the protection of unattended and highly sensitive information.

The use of AI in cyber defence systems serves as a hallmark for companies looking to enhance their security posture against increasingly complex threats. This case study highlights the development of AI in cybersecurity by documenting investment strategies in AI technologies and their resulting impact on the cyber defence ecosystem—shifting the paradigm from reactive to proactive, restoring agility and resilience for organisations operating in today's multifaceted world.

Weaknesses and Issues AI-Powered Defences

Without a doubt, artificial intelligence has changed the face of cyber defence by offering advanced capabilities in threat detection, incident response, and predictive analysis. Nevertheless, this innovation does pose some weaknesses and issues. One such issue is the possible exploitation of AI algorithms and models by malign cyber actors. Most AI systems are built upon the data input framework; therefore, adversaries may seek to compromise and poison the training data to bias AI defences by sabotaging the learning process.

Additionally, the complexity of AI algorithm features can give rise to possible weaknesses. An attacker could take advantage of subtle flaws in the application or design of AI systems to circumvent mechanisms intended for detection and to thwart the system's protective features. Automated decision making poses ethical and legal issues in all domains; for example, in domains where the outcomes, inaccuracies, or false, and over-predictions lead to damaging consequences.

Equally important is the issue of adversarial attacks tailored to more sophisticated and automated security systems; AI-offensive-defensive strategies which specifically target an organisation's algorithm-based security prepare "AI hunted" through a process referred to as perturbation and misclassification; causing an AI to incorrectly label items and evade detection. So, to function properly, a defender must reinforce their AI and adapt more sophisticated mitigation counteraction techniques against such adversarial movements.

A related issue is the shortage of experts in the two divisions of AI and cybersecurity; this lack of experts brings forth newly sophisticated problems. For advanced defensive systems driven by algorithms to function as they are meant to, there is a need for adept personnel; professionals who can build, execute, and supervise those advanced technologies. Harnessing this potential involves tackling the skill shortage and enhancing cross-disciplinary training.

The incorporation of AI into pre-existing security frameworks presents considerable difficulties. Older systems might not easily integrate with AI technologies, requiring major design modifications and possible disruptions during the shift. Moreover, the need for AI algorithms to be explainable and transparent is crucial, especially in heavily regulated sectors that necessitate responsibility and audit trails.

To sum up, AI can enhance and transform the capabilities an organisation has for cyber defence. In doing so, it can present new vulnerabilities and risks that require careful consideration and strategic approaches to remove or mitigate. Resolving these issues would allow organisations to strengthen their AI-based defences, while maximising the effectiveness of AI in protecting digital assets and privacy.

Building Offensive AI Capabilities

The development and use of offensive AI capabilities is a new area of concern in cyber warfare. As adversaries enhance their methods, defensive actors must do more than just strengthen their shields: anticipating, shaping, and actively countering emerging threats becomes a vital necessity. Constructing offensive AI capabilities is multidisciplinary. It combines predictive modelling, adversarial behaviour analysis, and intricate algorithms. It requires using AI to target system weaknesses, commandeer networks, and disrupt or mislead enemy operations. This work needs extensive knowledge of offensive cyber strategies and the ability to apply AI systems to optimise such strategies. The use of AI for offensive cyber operations profoundly raises ethical issues as well. It is critical to maintain ethical boundaries when developing offensive AI capabilities. Such conduct must be grounded in ethical and legal constraints which protect autonomy and order in cyberspace. Those who develop offensive AI capabilities must respect ethical obligations of transparency, responsibility, and compliance with applicable laws.

Moreover, the incorporation of offensive AI technologies within the broader context of cybersecurity requires regard for the potential consequences for international public order and security. This involves analysing the risk of escalation in conflict that might result from deploying offensive AI technologies and estimating their strategic value in relations between states. For responsible governance concerning risks and threats, the geopolitical context and far-reaching consequences of the use of offensive AIs in global cybersecurity systems are paramount for constructive action. Also, collaboration and the sharing of information among the stakeholders is crucial in facing the difficulties and dangers associated with the enhancement of offensive AI capabilities. For the comprehensive risks and effects of offensive AI in cyberspace, it is necessary to approach specialists such as in AI, cybersecurity, international law, and geopolitics. In the end, the application of offensive AI technologies requires profound reflection, systematic control, and serious implementation to strengthen cybersecurity systems, shield against digital aggression, and defend the norms of responsible state conduct in the international arena.

Strategic Consequences and Global Cybersecurity Developments

The advent of AI in the context of cyber warfare signifies a new chapter in the dynamics of global cybersecurity. Opportunities for both states and non-state actors to exploit AI offensively reshape the conventional threat matrix and, in turn, create a need for a more sophisticated strategic approach to counter such threats. This evolution has constellational implications, spanning across diplomacy, military affairs, technology, and beyond. On the global stage, the use of AI technologies for cyber offensives increases concerns related to the destabilising impact it may pose on existing security frameworks as well as the heightened risk of cyber conflict escalation. As a result, it is critical that decision-makers undertake coordinated negotiations to reach agreements aiming at setting standards and protocols for the employability of AI in cyber warfare.

Conceptually, the use of AI for offensive capabilities within a nation's military structure calls for an overhaul of the conceived doctrine and operational concepts. The new complexities of deterrence and defence brought by precision cyber strikes enabled through autonomous AI systems add another dimension to the rapid pace at which threats AI technologies present. Furthermore, the pace at which AI-driven threats evolve necessitates the constant advancement of defences, including the development of adaptive AI countermeasures and strategies designed to mitigate such threats.

Equally important is the technological component of this change, as it requires a rethink of architecture and protocols of cybersecurity. The increasing use of AI in offensive actions requires further development so that mechanisms for detection, attribution, and protection of critical infrastructure against sophisticated AI-driven assaults are strengthened. Also, the ethical and legal implications of using AI for offensive purposes create a gap that calls for policies without ambiguities and contradictions that would enable responsible, transparent governance while serving national security interests contrary to legal frameworks of sovereign states and human rights standards.

Taking into account the outlined issues, the active approach towards defence adaptation combined with resilience to systemic disruption must be adopted by organisational and state actors. A shift away from static defences places emphasis on active cyber defence, combining sophisticated AI-driven threat intelligence with agile counter-response capabilities, while focusing on capacity-building in AI-enhanced cyber defence technologies. The need for public-private partnerships to foster collective cyber resilience and protection of critical digital infrastructures against AI-enabled attacks becomes evident.

The more advanced AI technologies become, particularly regarding their use in cyber warfare, the more essential global cybersecurity issues are likely to require an in-depth analysis regarding countermeasures. Global cyber conflict is bound to change; an integrated approach to strategy development, guided by innovation in the utilisation of AI, will prove critical in safeguarding resilience, reducing risks, and protecting the cyberspace ecosystem.

2
Introducting AI's Role in Modern Cybersecurity

The Development of Cybersecurity Techniques

The advancement of digital technology spawned a nascent industry known as cybersecurity, which has markedly changed over the years. Emerging technologies such as computers, networks, and information storage systems advanced at a rapid pace, which, in turn, served to protect digital information. Indeed, there was a need to develop cybersecurity policies and protocols. The evolution of cybersecurity dates back to the time when computers first evolved. During this period, standalone (or single-user) computers were used for very basic data processing and storage. At that time, security measures were implemented mostly in the form of physical access control and basic user identification systems that allowed only permitted users to enter the network.

The introduction of the internet and the widespread adoption of digital technology revolutionised the interconnection of systems to a whole new level, giving rise to a new spectrum of cyber threats and weaknesses. Besides, with the increase in connectivity, so much sensitive data and information was targeted by malicious attackers. Highly sophisticated disreputable acts were on the rise accompanied by a transformative shift. That phase marked the development of modern cybersecurity techniques which included firewalls, antivirus programmes, as well as intrusion detection systems used to monitor the networks and alert users in case of security breaches.

The start of the 21st century marked a period of fierce cyber-attacks, which drove a change in cybersecurity strategies. Examples of these attacks incorporated the use of computer worms, denial-of-service attacks, and data breaches. Because outdated security methods could not handle these sophisticated attacks, professionals in cybersecurity were forced to shift their strategies. As a solution, IT experts began to implement more proactive and active defence strategies, including perimeter defence, continuous surveillance and real-time monitoring, incident management, and comprehensive vulnerability assessments.

Troublesome network weaknesses have always been a target for cyber criminals. As time progressed, the morphing of digital space gave rise to more complex and hidden network tactics. As a consequence, this prompted the birth of innovative cybersecurity paradigms, such as behavioural analytics and machine learning algorithms, as well as AI focused counter measure systems designed to predict, identify and eradicate ever evolving cyber threats. With the amalgamation of artificial intelligence and cybersecurity, organisations were able to improve their cyber defence. Access to advanced threat detection systems and algorithm driven, automated counter measures bolstered organisational security against sophisticated threats.

Moreover, the increasing interconnection of devices in the Internet of Things (IoT) and the increase in the number of cloud computing services attending to business needs created new problems that required a security solution to be as flexible and elastic as the modern IT system itself. As a result, the change in cybersecurity practices has served as a testament to the unyielding spirit and adaptability of security practitioners towards relentless shifts in the digital world, continuously adapting to persistent attempts from malicious forces.

AI as a Transformational Force in Digital Defence

The remarkable, rapid progression in our world's technology ecosystem offers unimagined possibilities; however, its unprecedented advancement makes lives easier, it also creates professionally sophisticated challenges to the businesses. The use of artificial intelligence (AI) has quickly become one of the central pillars to digital defence architecture, offering remarkable prospects in prevention, detection, prediction, and reaction capabilities against cyber-attacks.

Understanding Cyber Threat Landscape Through AI Eyes

The new realm of modern cyber threats has opened

new avenues of exploration. It is clear that the age-old approaches to threat detection and mitigation may not work with the modern threats. The nature of cyber attacks requires organisations to be proactive and agile in their approach towards cybersecurity frameworks. AI is the technology that provides a paradigm shift in digital defence. By employing AI technologies, organisations can understand the cyber threat landscape meticulously, which will enable organisations to anticipate and neutralise latent threats with unmatched accuracy and efficiency. AI gives, in this context, a new dimension to the challenges facing the security professionals owing to its ability to analyse huge datasets in real time, and identify important trends and relationships between various targets of compromise which would otherwise go unnoticed.

AI continuously learns and adapts to its environment, which helps in the analysis of new vectors emerging along with surviving older ones, thus enabling a well-rounded preview of the cyber climate. With the use of sophisticated algorithms and the learning ability of computers, AI is capable of identifying sophisticated patterns of the attack and responding appropriately to universal and exceptional pivotal avenues rapidly to the defences and thus lessen the risks exponentially.

This knowledge affords security teams a considerable benefit when contending with advanced cyber threats that use stealthy and multi-layered approaches. Additionally, the ability of AI to forecast upcoming scenarios enhances the readiness and resilience of security practitioners by enabling them to prepare for various attack vectors. From AI's perspective, the cyber threat landscape metamorphoses from an opaque, daunting expanse into one that is analyzable and, ultimately, manageable. Rather than facing an inundation of data, actionable insights can be harnessed, and responses tailored to the AI-assisted context. That being said, AI gives unprecedented visibility into the threat landscape, but this is most effective when there is a collaboration between humans and machines—the interplay of human intelligence and AI-powered insights. As with all technological systems, there is a degree of output interpretation that requires human intuition, critical thinking, and expertise in the field to guide strategy and situate anomalies, which AI systems identify, in the extensive organisational frame. Within this structure, AI empowers cybersecurity specialists by taking over routine activities and improving the overall posture of the system. Thus, cyber adversaries can be anticipated, monitored, and preemptive automated countermeasures triggered with minimal human supervision.

Overview of AI Technologies in Use Today

Modern AI technologies have become critical components of the contemporary approaches to cybersecurity. The advancement of cyber defence capabilities and active risk mitigation is made possible through the use of sophisticated algorithms, AI, machine learning, and large data sets. In the AI domain, AI is being applied for multiple purposes in cybersecurity, including but not limited to, threat detection, anomalies detection, behaviour monitoring and evaluation, and pattern recognition. AI technology-based security solutions utilise neural networks, deep learning, natural language processing, and predictive analytics. With these, organisations are able to advance their security posture against sophisticated digital attacks. An inordinate effect on the operationalisation of AI technologies in cybersecurity is due to Machine learning technologies. It enhances the extraction of actionable intelligence from large datasets through some level of automation, improving the precision and timeliness of detection and response to the threats.

Machine learning provides automated, responsive, adaptive features to security systems such that the given level of security can improve/set standards based on continuous observations or feedback systems. Furthermore, AI technologies are incorporated into Cybersecurity systems with the purpose of automating non-strategic routine tasks thus reducing the workload for human operators. Through the use of AI systems, organisations can automate with ease semi-automated log-analysis, automatic vulnerability detection, automatic incident response, and other processes.

AI enhances these tasks with contextual insight, which aids in decision-making and reduces the time needed to respond. Also, cognitive computing, which is a form of

AI, has now become very useful in examining complex security situations, as it attempts to think like humans. This assists in a more nuanced assessment of complex threats and, more importantly, the recognition of attack patterns that have not been previously detected. A rapidly expanding area of interest is combining AI with threat intelligence platforms, which can process multiple streams of data to generate actionable insights. With AI, organisations can swiftly distil threat vectors and assess in real-time which require immediate action and strategised mitigation. Advanced AI technologies will enable stronger cybersecurity architectures with predictive analytics and proactive defence strategies. Cybersecurity practitioners need to remain aware of new developments in AI and incorporate its new advances responsibly into frameworks in order to mitigate new risks and protect assets.

Comparative Analysis: Traditional vs. AI-Enhanced Security

Input Volume Cyber attacks Cyber security breached Instance Detonated, traditional security resorted to distinct measures with log systems in place, responding manually to threats by citing evidence, taking actions based on frameworks and reliant on evidence systems. Such strategies, while somewhat manageable in the past, are quickly becoming outdated and ineffective due to the volume, variety, and skill level of modern cyber attacks. On the other hand, Real Time Diagnostics Systems AI-powered security solutions focus on monitoring using advanced AI tools such as machine learning, natural language processing, and even behavioural analytics, identifying patterns, anomalies and threats in real time, going far beyond the limits of pre-conditioned rules and signatures. The transition from inflexible, preconditioned security systems based and focused on rules to self-adapting systems has provided a pathway towards autonomous cyber threat prevention.

Machine Learning Enhanced Search distinguishes itself uniquely in tackling unseen threats compared to traditional AI and Machine Learning; deep learning is able to examine historical data to devise and detect new attacks infiltrating through masked systems. Zero-day attacks are advanced and novel techniques employed by sophisticated security actors which are self-sufficient in capturing and introducing modern security tools. Multi-layered rules targeting the conventional security set are unable to detect advanced attacks and struggle to TRA algorithms that can detect malicious patterns require pre-defined parameters to work and thus fail even when heuristic systems are employed to catch novel attempts from advanced participants. The ability both to continuously evolve and learn new data provides AI security solutions with a decisive edge over static, rule-based security measures.

Beyond this, a multitude of traditional security measures have been known to cross-check large databases with disproportionate amounts of irrelevant information, as well as ignoring the core aspects of a given situation while expounding on meaningless details. All of this serves to make identifying actual threats daunting. Comparatively, security measures powered by artificial intelligence do the exact opposite. These systems enhanced with AI technologies prove to be far more efficient in distinguishing real threats with patterns formed from the correlation of shifting data. Through these channels, and as a by-product of the technology streamlining responses to events, professionals can instead prioritise addressing real security events rather than bearing the burden of false alarms that the Xai models throw in their face, which indeed require the uncritical sifting through enormous heaps of erroneous information. The intricacy of these processes is far more precise than anything a mere human can administer.

The effectiveness of addressing any problem gets significantly affected by the speed and accuracy of equipping any response to it; therefore, identifying how fast tensions are able to sense triggers that activate every sequence that they have preset, as well as defining each line that should be crossed while switching from "neutral" to "alarming" should be observed too. Just as humans get mentally tired combing through vast streams of logs trying to spot a breach in the security loophole, there is always a parallel AI willing to do that for them. This does not need to take a long time; indeed, the extension of time provided serves for calculated enhanced strategies.

Security systems powered by AI technologies can analyse enormous datasets of logs and documents in order to uncover potential threats far more efficiently and on-demand than humans, sending real-time alerts of emerging new ones or even neutralising certain assaults without any human intervention. Put differently, more intelligent responsiveness pairs bounded automation of low-level countermeasures free up functions, which markedly diminishes the quantum damage done to an organisation from hours and days of waiting and suffering instead of minutes and seconds.

To conclude, the move from conventional to AI-integrated security systems marks a critical turning point in cyber-defensive strategies. With AI, organisations can now dynamically adapt to change, compelling cyber threat detection, minimising incorrectly flagged risks ('false positives'), and counteracting emerging risks in real-time.

Key Benefits of AI Integration in Cyber Strategies

The integration of Artificial Intelligence (AI) in cyber strategies offers myriad benefits that bolster the protective and responsive capabilities of security measures in an organization. One of the primary advantages is real-time threat evaluation and analysis, which is possible due to the speed at which AI tools process data.

This proactive stance empowers organizations to avert security breaches and possible cyber-attacks, which reduces the exposure of the firm to losses. AI also adapts remarkably well to recognising patterns or anomalies, and therefore, can identify many new or sophisticated attack vectors that have previously gone unrecognised.

Furthermore, the use of AI in cybersecurity makes it possible to automate monotonous workflows, which lessens the load on personnel and enables security teams to tackle strategic planning. The reduction of manual intervention not only optimises productivity, but also reduces the range of errors and oversights. AI enhances the agility of security operations by automating repetitive tasks, thus improving the speed of incident response and resolution. AI can automate repetitive tasks such as patch management, log analysis, and other defined procedures. Such acceleration fortifies the organisation's security posture.

Moreover, AI's capability to apply its self-performed duties and functions through the use of multiple machine learning algorithms enables it to improve its performance and efficacy over time. It is capable of iterating learning and performing analyses, therefore refining its understanding of a network's normal behaviour over time. This ability enables a stronger distinction amongst genuine anomalies as opposed to false positives. This facet of self-improvement enabled through AI fosters a proactive line of defence that is adaptive alongside the ceaselessly shifting threat environment.

The integration of AI comes with numerous benefits, one of which is the ability to predict possible vulnerabilities, as well as future scenarios of attacks based on historical information and emerging trends. The organisation fortifying security gaps before other entities are able to take advantage of them enables preemptive risk mitigation, therefore protecting critical infrastructure from full-fledged threats. The framework employing this approach enhances the protective capability and resilience of infra systems, thus minimising the chances of successful attacks using cyber means.

To conclude with some of AI's most salient attributes in cyber strategy, real-time threat detection and anomaly recognition, automation of responsibilities and tasks, continual improvement, and even predictive analytics stand out the most. Organisations able to embrace these advantages are sure to bolster their cybersecurity posture and defences toward an entire menu of digitally evolving threats.

Addressing Challenges: Implementing AI in Security

The deployment of artificial intelligence (AI) in cybersecurity comes with numerous challenges that can limit how an organisation seeks to exploit its capabilities fully.

One of the basic challenges is the lack of available talent and expertise for AI technologies in the field of security. Obtaining skilled personnel capable of building, operating, and maintaining AI-based cybersecurity systems is proven difficult, owing to the highly specialised nature of the field. There is also a need to restructure the available workforce in security, which will involve training them to work with AI technologies, further straining the available resources. From a different perspective, there are challenges of moral responsibility attributed to the application of AI in security. The misuse of AI-based instruments or algorithms can pose serious ethical challenges that call for very strict policies to avoid these impacts and control problems.

The rapid pace of the cyber world sophisticated crimes means that AI models must be agile, agile, and frequently revised in new ways to outsmart criminals. That flexible aspect makes it imperative that there is always R&D directed towards ensuring adaptation for new challenges. There are also problems that concern lack of integration or interoperability of security systems with AI as other AI-based segments. Most organisations already possess complex legacy security ecosystems consisting of heterogeneous tools and appliances.

AI's integration into the pre-existing structures requires careful configuration and implementation strategies to ensure that there is seamless interoperability between the security frameworks AI technologies integrate with. Moreover, AI's decisions in relation to security systems present a challenge concerning explainability. Trust and accountability necessitate understanding the rationale behind specific actions attributed to algorithms, especially when they reach conclusions. This emphasises the need for creating adoptable and acceptable AI systems – transparent and interpretable systems – fostering trust. Issues of compliance and regulations add a layer of complexity in relation to the cybersecurity domain. Adopting sector-specific guidelines, global benchmarks, and data protection frameworks creates difficulties when implementing AI technologies in the cybersecurity field. Addressing these matters necessitates collaboration between diverse security practitioners, AI practitioners, compliance and regulatory agencies, and industrial representatives. Long-term collaborative partnerships characterised by strategic human capital, solid governance, unending inquiry, and proactive interdisciplinary integration stand to effectively resolve the challenges posed by the AI-cybersecurity integration puzzle.

Case Examples of AI in Action

Examples of AI applications in cybersecurity are becoming increasingly prevalent. Noteworthy applications include the use of machine learning algorithms for anomaly detection and the counteraction of advanced persistent threats (APTs) in enterprise networks. Businesses can proactively neutralise potential breaches by mitigating anomalous behaviour patterns identified by

AI and learning from attacks that continuously evolve over time. Moreover, AI anomaly detection systems significantly improve threat detection by decreasing false positive alerts.

One other very striking example is AI's ever-growing ability to examine vast amounts of data in order to find hidden "gems" or security loopholes. A case in point is when an AI system vetted millions of lines of code and scanned it for bugs, shortening a process that would on average take weeks to identify weaknesses, down to hours. Additionally, AI-augmented threat intelligence systems are critical in compiling diverse data sets to provide insightful and groundbreaking concepts on emerging cyber threats which in turn help security personnel gain the upper hand over their rivals.

Furthermore, AI's application in dynamic risk-based authentication has transformed access control systems, successfully preventing unauthorised attempts at access. The automatic behaviour-based models of authentication which evaluate users and contextual information provide organisations with the ability to not only considerably lower breaches of identity, but also reduce infringement on usability. In the same way, identity and access management systems powered by AI have simplified the management of privileges while guaranteeing proper rights, and access control has been enforced over a wide range of users and systems.

AI has showcased its ability to contain and remediate

security incidents with little human input which exemplifies its role in automating response to incidents. AI has managed cyber incidents by orchestrating and executing predetermined countermeasures proportional to real-time threats, thereby reducing the cyber incidents' impact, exposure duration, and recovery time. Such responsiveness has helped eliminate possible downtimes and avert financial losses.

These particular examples highlight the increasing use of AI as a powerful tool in bolstering cybersecurity rather than enforcing traditional methods of shielding systems. These examples demonstrate that the adoption of AI technologies enables organisations to effortlessly not only modernise their approach to cyber resilience, but also underscores the transformative impact of AI integration on organisational agility.

Fostering Full Adoption of AIs in Organisations

As discussed in the previous chapters, the application of artificial intelligence (AI) in cybersecurity practices has proved to be revolutionary and critical. Undoubtedly, AI is here to stay and it is not merely an industry catchphrase or a passing fad. In fact, AI is emerging as a vital weapon in combating cyber threats. Organisations are still unable to realise the full benefits of AI in strengthening their cyber defence systems because there are no comprehensive strategies developed and experts expect fully integrated frameworks to be employed by organisations.

Understanding the integrated framework to be employed by an organisation is the first step in effective AI implementation. Organisational cybersecurity challenges must also be addressed, along with aligning operational needs and strategic objectives. Focusing on the established security infrastructure, addressing the weaknesses of inflexible systems, and identifying possible gains from AI are critical concerns for decision-makers.

An initiative central to fostering the adoption of AI in cybersecurity is to build an adequate level of technical expertise internally. This involves initiating an AI-aware culture, facilitating collaborative work between cybersecurity experts and data science specialists, and recruiting training programmes which adequately prepare personnel to adopt AI technologies. Furthermore, organisations need to establish frameworks for continual AI learning and adaptation in cybersecurity due to its constantly changing nature.

Simultaneously, intertwining AI with pre-existing security workflows and processes is equally important. Cooperation between information technology departments, security operations, and AI application specialists is crucial for AI solutions to work synergistically with existing security architectures. Besides, the frameworks emphasise interoperability and scalability that will enable organisations to deal with changing threats and new technologies while still providing strong security.

Along with the technical aspects, the ethical and regulatory concerns of AI technologies also require attention in detail. There are obvious ethical, legal, and regulatory frameworks when it comes to AI Powered Cybersecurity Operations, especially as perception-driven AI systems are put into place. Designed systems that guarantee privacy, transparency, accountability, reliability, and integrity must conform not only to industry norms but also to existing regulations.

The integration of AI technologies should not be attempted by organisations only once; it is a journey where the focus should be on enabling as much AI potential as possible while avoiding the risks that accompany them. Through clearly defined paths for AI implementation, capacity building, ethical frameworks, environmental shifts, and compliance adaptability, organisations will be well-positioned towards the adoption of comprehensive AI in contemporary cybersecurity and bolster defences from incessant digital assaults.

Conclusion and Further Discussion

Due to the growing fragmented nature of hazardous activities and the sophistication of information technologies, it is apparent that modern cybersecurity AI solutions are not only favourable but utterly indispensable. The soaring and descending technologies of AI have moulded the traditional methods employed in cyber defence and offence, giving rise to new and more dynamic approaches to be taken. The purpose of this book is to provide the foundational knowledge, applications, and implications of AI in cybersecurity in order to arm readers with sufficient understanding of this evolving sphere. As we reach the end of this discussion, we must understand the focus for now, which is the scope of changes anticipated in the future.

Anticipating where we left off, the industrial security systems incorporated with AI features will create a more complex world in the scope of safeguarding cyberspaces. With advanced persistent threats and an ever-growing list of attack vectors, organisations with proper defences can leverage machine learning algorithms, natural language processing, as well as predictive analytics. Further, the responding systems to incidents will experience a paradigm shift for the better as the now-independent systems powered by AI will virtually change how these security incidents are addressed in the real world: swiftly and accurately. Given the right structure, organisations will have eagle eyes on their cyberspaces, and risks can be proactively neutralised thanks to AI's ability to process real-time data.

The AI and cybersecurity ethics, legal frameworks, and policies will be critical in determining the future of the industry. The evolution of AI technologies creates stronger needs for governance policies with responsible thresholds set using these technologies. There needs to be collaboration between lawmakers, business executives, and cybersecurity experts to determine the standards and policies that guarantee privacy, transparency, and reasonable treatment in AI-powered security processes.

Moreover, the integration of quantum computing, blockchain, IoT, and other revolutionary technologies with AI will create endless possibilities and complications. The combination of these technologies can radically change the physics of cyber technology while creating additional security challenges as well as new ways to defend against attacks. With the advancement of AI technologies, the application of adversarial machine learning and self-learning AI to defeat sophisticated assaults will demonstrate the extent of changes in cyber resiliency.

To summarise, the pursuit of fully integrating AI into cybersecurity is a work in progress with perpetual refinement and ingenuity. The advancing forefront of technology—which encompasses the use of AI—articulates the need for a meticulous blend of virtue and ethics in cybersecurity. As the landscape of digital threats broadens, so does the importance of AI technology to provide virtual protection. Such fortifications of our cyber infrastructure are essential in aiding the industry move boldly towards a future unencumbered by the suffocating grasp of cyber threats.

Foundations of AI in Cyber Defense

3

The Evolution and Historical Context of AI in Cyber Defence

The use of AI in cybersecurity today is an advancement that has evolved due to various historical events and advancements. One of the early concepts of machine learning and its application to security systems serves as an initial thinking point for this framework. One major landmark in history is the invention of neural networks, from which more sophisticated data analysis and pattern recognition systems were developed. Advanced, dynamic, and intelligent defence systems were and still are fundamentally essential, especially when taking into consideration the ever-changing cyberspace threat landscape. The persistent integration of AI solutions into cybersecurity systems reflects this need. The digital transformation alongside the growing complexity and volume of cyber-attacks demonstrates the historical need for having AI in cyber defence and utilising it as a powerful tool. Other historical events like major cyber-attacks or data breaches have called attention to the need for stronger defences. These instances, combined with the progression of technology, showcase the continuous evolution of the relationship between AI and cyber defence.

With the progressing impact of AI technology, novel advancements, such as forecasting and self-initiating attack recognition systems, have emerged within the scope of cybersecurity. These advancements have revolutionised the domain of cyber defence, enabling organisations to proactively counter developing threats. Understanding the history of AI in cyber defence helps us appreciate and understand the evolution that led to today's use of AI as the pillar of modern cyber defence techniques and strategies. Studying the historical development of AI in cyber defence would help us better comprehend the defining milestones and the AI-driven shifts that prompted embedded resilient and adaptive security systems frameworks.

Key Concepts: Machine Learning, Neural Networks, and Beyond

The change brought about by AI in cybersecurity industry processes such as threat detection response and classification is made possible by Machine Learning (ML). In cyber operations, ML algorithms are implemented for anomaly detection and predictive analysis for security lapses. Neural Networks, a branch of ML, mimics human thought processes for information processing through multi-layered and linked algorithms, enabling sophisticated decisions. The combination of ML and Neural Networks allows for the creation and continuous improvement of advanced adaptive and self-learning cybersecurity solutions.

Another primary AI concept, Reinforcement Learning,

makes it possible to train autonomous agents to make decisions in succession and under uncertainty, which is invaluable to cybersecurity domains that require rapid responsiveness. AI also includes Natural Language Processing (NLP), which provides the means to process and analyse huge volumes of unstructured data like texts, security logs, incident reports, and threat intelligence feeds. NLP helps in automating the analytic function of cybersecurity specialists in the interpretation and response to emerging issues by providing actionable information derived from text data.

The recent emergence of Generative Adversarial Networks (GANs) is one of the newest advancements in AI that creates synthetic data for training and evaluating cyber defence systems; thus this approach greatly aids in improving the variety and resilience of the defensive strategies used. Cybersecurity measures are traditionally reactive, but artificial intelligence as a whole is transforming the field by assisting in cyber defence capabilities with the ability to identify and address intricate and evolving threats, thereby redefining the strategy used to protect digital property and infrastructure.

Core Principles of Cybersecurity Enhanced by AI

With the rapid expansion of technology, cybersecurity has developed into a complicated yet essential facet for businesses and other modern-day operations. Today, due to the ubiquity of information systems, the use of new technology such as Artificial Intelligence (AI) is proving to be significantly helpful in strengthening protective measures against cyber threats. One notable area that is changing the central tenets of cybersecurity is the functionality of AI, which is significantly altering the methods used by enterprises for threat identification, prevention, and response.

Proactive threat identification is one of the AI-enabled advancements in cybersecurity. Most cybersecurity frameworks employ a threat identification and mitigation technique after an attack has already taken place. AI propels systems to perform real-time analysis of data streams, detect possible breaches and vulnerabilities long before they occur. AI is capable of domain-independent anomaly detection and anomalies through the application of its advanced algorithms, providing organisations the power to undertake preemptive defensive actions on their data and systems which requires great vigilance.

AI has also made it possible for organisations to intelligently set up adaptive and agile defenses. The gradual evolution of cyber threats calls for a flexible approach to cybersecurity and elevates the need for AI-integrated systems. Constant scrutiny of an organisation's computer networks enables AI to monitor, analyse and swiftly reposition its own defence systems to counter attacks, therefore reinforcing the organisation's cybersecurity strategies.

AI enhances traditional security mechanisms through the use of predictive analytics. With the increasing number of predictive algorithms and historical data, AI is able to generate more powerful insights that predict forthcoming security threats and risks, therefore enabling organisations to prepare for potential approached threats.

Moreover, AI helps automate routine security processes, which greatly improves operational productivity. Through automation of patch management and response to security incidents, AI not only streamlines these tasks, but also enhances operational efficiency, allowing human personnel to dedicate their efforts toward strategic tasks. Firms can strengthen cyber defenses and improve resource allocation within their cybersecurity frameworks by automating such systems.

Furthermore, AI actively enhances the cybersecurity orchestration processes. AI not only allows for the integration of various security systems and technologies but also aids in the coordination and correlation of security events and alerts throughout the organisation's infrastructure. This capability aids in the enhancement of visibility and control and thus, allows security teams to obtain detailed intelligence and effectively manage strategic responses to incidents.

All in all, AI's advance integration into cyber defence strategies upholds the tenets of confidentiality, integrity, and availability. This allows organisations to better manage the confidentiality of sensitive data, safeguard systems, and ensure vital services are consistently available. With innovations in technology, the repercussions of AI are bound to shift the principles and practices employed in defending systems from digital assaults.

AI Algorithms and Models: A Technical Overview

While tackling cyber threats, knowing the relevant algorithms and models in AI is an absolute necessity. Cybersecurity encompasses a broader spectrum of challenges that require different approaches because of the continuously changing landscape of threats. At the heart of all new developments, there lies a branch of AI known as machine learning. In this context, different types of learning algorithms, including supervised, unsupervised, and reinforcement learning, are crucial. Prediction-based learning using labelled datasets defines supervised learning, pattern recognition in unlabeled datasets defines unsupervised learning, and agent training for sequential decisions using trial-error methods defines reinforcement learning. Most entailments of AI solutions are based on these principles.

Neural networks such as deep learning models are becoming more popular for a wide range of applications, including but not limited to, malware detection, network intrusion detection, and anomaly detection. The application of deep learning to multidimensional complex data in cybersecurity has proven to be very effective by using CNNs and RNNs. Also, ensemble methods which improve prediction accuracy by combining several models, and some techniques from NLP are now being used more often within AI to better counteract cyber-attacks.

Coalescing AI with graph theory and complex network analysis goes beyond extraordinary algorithms and offers new ways of understanding interrelated dangers and possible attack pathways. This blending fosters a comprehensive grasp of the cyber ecospheres and, in turn, more effective protective structures. The evolution of AI has also transitioned interest towards XAI or explainable AI aimed at elucidating the processes undertaken by AI models towards cybersecurity decisions owing to trust and transparency quandaries. The systemic study of AI models and algorithms affirms the diverse and complex character of cyber defence AI and speaks to its revolutionary capability in enhancing the global digital defence architecture.

Data-Driven Insights: Harnessing Big Data for Cybersecurity

The digital evolution of the cybersecurity industry has sparked the emergence of new technologies capable of sophisticated advanced analytics to address the ever-growing challenge of processing large amounts of data. Here, we address the importance of big data in enhancing cyber defence mechanisms and covers how insights derived from data analytics can transform the practice of cybersecurity. Protecting against cyber threats entails application of big data, which requires the safeguarding and storage of information, as well as the extraction of intelligence from the data. Utilising sophisticated data overflow analytics, organisations can discern concealed structures, anomalies, and relationships, resulting in deep visibility into network activity. Additionally, big data enables the cybersecurity professional to identify possible threats and move proactively to mitigate emerging attack vectors.

Our focus here is on the use of big data for threat hunting and for responding to or investigating incidents to emphasise the need for improving cyber resilience. The combination of machine learning with big data enables automated anomaly detection, predictive analytics, real time risk assessment, and response to dynamic cyber threats instantaneously with utmost precision.

The application of big data in cybersecurity is certainly not a simple task. Meeting the velocity, variety, and veracity of the data demands unprecedented infrastructure, complex architecture, and high-level governance processes. On the other hand, privacy and compliance issues also require careful safeguarding of sensitive information while utilising big data in cybersecurity. In solving all of these issues, it is important for companies to observe legal restrictions and ethical principles concerning the big data undertakings. With the proper implementation of big data, cybersecurity experts can augment their defending capabilities, outsmart their adversaries, and fortify their systems against sophisticated attacks. More than ever, instead of focusing on the security measures, organisations turn to the defensive quadrant powered by intelligence in the fight against cyber threats which emphasises the critical importance of these insights.

The Role of Anomaly Detection in Threat Identification

Within the realm of contemporary cybersecurity, anomaly detection is essential for recognising irregular activities in a system or network. As the threat landscape becomes more dynamic and sophisticated, signature-based or rule-based detection approaches become increasingly obsolete and incapable of identifying new attacks or complex tactics employed by adversaries. AI and machine learning-powered anomaly detection systems are the most efficient at this task. Anomaly detection algorithms have the potential to uncover bad behaviour by spotting deviations from normal activities after examining copious amounts of data. Anomaly detection's flexibility to respond to shifts within the threat landscape and not permanently rely on rules or signatures is one of its most powerful assets. This adaptability greatly enhances the capability for proactive threat detection. Integrated with AI-powered cybersecurity frameworks, anomaly detection arms organisations with critical capabilities for dealing with new and unknown threats. Although, the need to accurately tell apart genuine benign variations from genuine anomalies, high false positive rates, real-time detection without system lag, and zero impact on system performance all make implementing effective anomaly detection challenging. Advanced intervention and constant model refinement is needed to surpass these obstacles—and there is a blend of heuristic approaches, robust model training, and dynamic system tuning required to arrive at optimal performance.

Methods for anomaly detection differ from one another as they may include statistical and clustering techniques, as well as approaches based on deep learning, which all have their pros and cons. Organisations must conduct a proper analysis for the techniques being considered to ensure that none of them are blindly ignored and that the cybersecurity measures required are tailored and personalised.

Anomalies and models of detection require constant adjustment and supervision to catch up with the new, more sophisticated threats and to uncover the new pathways to attacks in time. In essence, the contribution of anomaly detection in threat identification does not end at risks; adaptive smart cyber defence systems that allow the defender to anticipate and intelligently respond to adversarial moves with greater speed, flexibility and accuracy rely heavily on automated counters.

Challenges in Implementing AI for Cyber Defence

The opportunities that derive from artificial intelligence's (AI's) application to cyber defence come along with numerous hurdles that require proper consideration and planning, the most pressing being the constantly changing and Advanced Persistent Threats (APTs) which require AI solutions that are both predictive and responsive to new challenges. Furthermore, the requirement of operational AI in conjunction with already existing cyber defence systems presents a considerable challenge due to the concerns of flexibility and adaptability, sustained reliability, and operational consistency in performance across different ecosystems. Also, ethical dilemmas regarding the AI resolutions in matters pertaining to critical issues of cyber defence posture fusion complicate the human-centred governance approach where responsibility and control rests heavily with people.

The inadequacy of annotated datasets to train cyber defence AI models greatly inhibits achieving the needed underlying accuracy and amplifies the demand for improved data collection and annotation solutions. Finally, gaps in AI system neural network models which might result from targeted attacks intended to subvert the efficacy of AI powered cyber defence system tools require concrete and proactive counteraction.

The growing intricacy regarding the implementation of AI in cyber defence systems deepens functional and resource requirements, forcing organisations to confront multidisciplinary issues such as skill gaps, acquiring cybersecurity personnel, and the funding needed to ensure effective AI adoption. The need to balance timely threat detection with response action and minimisation of defence-shield breaches and countermeasure false alarms introduces yet another layer of intricate challenge requiring extensive tuning and validation. Finally, relevant regulations and compliance policies governing AI technologies in cyber defence solutions add meaningful complexities pertaining to alignment with legal frameworks, governance, and social responsibilities on technologies and practices employing AI. To solve these interconnected problems, there is a need to incorporate alternative approaches including technological solutions, systematic and cross-disciplinary stakeholder collaboration, and agile approaches accommodating shifting parameters in the cyber world.

Integration of AI with Existing Cyber Defense Frameworks

Incorporating Artificial Intelligence (AI) into existing frameworks of cyber defence systems is a drastic change in how organisations defend against sophisticated cyber attacks. Adversaries cause unprecedented challenges to cybersecurity which stem from how quickly the world is going digital. This invites the need for advanced technology, particularly at the suggested level with AI. Integration of AI with existing cyber defence frameworks differs at the strategic level. To begin with, organisational cybersecurity architecture needs to be evaluated to understand where AI could add value, thus creating a model of requirements aimed solely at identifying gaps for maximising the potential role of AI. The assessment involves reviewing all pertinent security measures, network topology, and threat detection systems to identify peripheral systems with suitable AI. After this adjustment, a customised tactical plan that reinforces systems with AI-powered tools is formulated. This plan includes the application of machine learning algorithms for real-time threat identification, predictive analytics for proactive risk identification, and natural language processing for data source evaluation of unstructured information.

Moreover, the merger process includes the seamless merging of AI-enabled systems into existing security infrastructures, ensuring that operational continuity is sustained throughout the entire integration process. Further, integration entails instruction and career advancement related to AI technologies for cybersecurity professionals, which in turn helps refine integration practices. This equips the team with the necessary knowledge and skills to employ emerging AI tools and effectively adapt to the changing security environment. Furthermore, as part of the integration process, persistent policies for governance and compliance regarding the application of AI in cyber defence should be designed. This entails designing frameworks for ethical purposes regarding the means of employing AI solutions, safeguarding data privacy and protection, and ensuring compliance with pertinent regulations and prevailing standards. Another distinguishing characteristic is the constant scrutiny and overhaul of the integrated AI cyber tools' functionalities within the scope of set cyber defence systems. This calls for the establishment of feedback loops and metrics to assess performance relative to AI-based defence mechanisms. Through audits and assessments, organisations identify gaps that require attention to enable agile refinement of AI-enhanced systems.

The successful integration of AI with existing cyber defence frameworks enhances the efficacy of proactive and adaptive defence postures, allowing organisations to become increasingly agile against changing cyber threats. Leveraging AI capabilities in threat detection, intelligence analysis, and response automation helps organisations bolster resilience to sophisticated aggressors while protecting digital assets and operations.

Technological Enablers: Hardware and Software Innovations

Progress is being made in developing technological

enablers as the domain of AI in cyber defence evolves, particularly in the realm of hardware and software innovations. The adoption of specialised hardware such as GPUs and TPUs is enabling faster analysis of enormous datasets by making the training and inference processes of AI models more efficient. Equally important is the promise that quantum computing holds: it will dramatically enhance the capability of AI systems with respect to cryptographic algorithms and code-breaking techniques, thus presenting both opportunities and challenges in cyber defence.

Cybersecurity specialists can now build advanced systems powered by AI for threat detection, anomaly detection, and behavioural analytics thanks to the development of AI-based solutions frameworks and libraries like TensorFlow, PyTorch, and Scikit-learn. Furthermore, the application of containerisation technologies like Docker and Kubernetes has made it easier to deploy and scale the functionalities of AI-powered cyber defence systems, thus optimising operational efficiency and resource utilisation.

Edge computing and 5G infrastructure allow AI systems to process data and make decisions at the network's edge in real-time, improving responsiveness and agility in mitigating cyber threats. Coupled with advances in distributed computing and federated learning, these technologies enable organisations to apply a collective intelligence paradigm in drawing insights from various data sets while preserving privacy and compliance with data legislation.

The adoption of AI-based security SOAR platforms automated most functions associated with responding to incidents while providing new methods for threat containment, remediation, and recovery procedures. This phenomenon has transformed incident response capabilities in cybersecurity. These platforms utilise AI to perform advanced alert correlation, automate unsophisticated reaction processes, and work with the current security architecture to enable the cyber defence team to respond to highly sophisticated and ever-evolving cyber warfare with speed and accuracy.

In the future, the combination of AI with new technologies like blockchain, homomorphic encryption, and secure multi-party computation might improve the effectiveness and secrecy of AI-powered cyber-attack prevention systems. With these technological enablers, the international community will be able to strengthen its protective mechanisms and face new challenges with more proactive, adaptive, and stronger cyber defence strategies.

Case Examples: Successful AI Applications in Cyber Defence

The field of cyber defence has witnessed an evolution

in the use of artificial intelligence (AI), with the focus now being on protecting critical infrastructure and sensitive data. Selected case examples help illustrate the role of AI in strengthening defences.

A vivid example is the implementation of anomaly detection AIs in the security system of a financial institution. The system, through the applied machine learning processes, autonomously recognised abnormal activities in the Network Traffic, consequently detecting advanced attempts of intrusion much quicker than traditional security frameworks could.

Another example can be seen in predictive intelligence for dealing with threats. A transnational company leveraged both AI and big data, allowing it to detect emergent cyber threats and take mitigative actions before disruption to operations occurred.

The autonomous incident response systems powered by AI have dramatically improved cyber defence capabilities. During a government agency cyber attack simulation, an AI-powered response system intruding on the agency's network automatically isolated infected systems, contained the malware infections, and started cleaning processes to fast recover systems, minimising damage and restoring normal operations quickly.

Moreover, the use of AI in penetration testing has transformed vulnerability assessment processes. AI tools were able to credibly attack organisational networks to find weaknesses, and thus, AI tools uncovered critical vulnerabilities that needed to be fixed immediately to increase the chances of being defended against before genuine attackers used these vulnerabilities.

This serves as an effective showcase of how modern AI technology can mitigate cyber socio-technical adversities. At the same time, however, this indicates that continuous innovations in AI technologies are needed to address constant changes in cyber technologies and the threats they pose.

4
Building Intelligent Defense Systems

Introducing Intelligent Defence Architectures

The integration of advanced technologies like AI, machine learning, and predictive analytics into intelligent defence architectures transforms cybersecurity by strengthening digital infrastructures against sophisticated cyber attacks. These architectures approach cybersecurity proactively. Organisations can now predict and adapt instead of reacting using previously implemented measures. The backbone of intelligent defence architectures is the application of AI, which enables systems to autonomously detect, assess, and respond to potentially harmful events in real time. Organisations can better defend themselves from these attacks by leveraging AI, enabling them to respond to geopolitical shifts and radical changes within the cyber threat environment.

The newest developments in intelligent defence architectures change how security teams view and mitigate risks. Infused AI into defence systems allows for advanced threat detection and mitigation, including dealing with high volumes of data and complex patterns. These architectures are capable of learning from new data at any time and adapting, making them more resilient to emerging threats and ensuring robust cybersecurity postures. Organisations' operational efficiency improves with dynamic, multi-layered adaptive response capabilities and reduced impact from severe security breaches.

In the framework of intelligent defence architectures, system components are integrated to create a cohesive multi-faceted defence structure. From behavioural analysis and anomaly detection systems, each building block contributes to a sophisticated layered defence approach that proficiently uncovers both latent and overt dangers. Moreover, intelligent defence architectures highlight situational awareness. Context-aware security systems have the ability to assess security events, allowing for a more prioritised and graded response relative to the importance of the involved assets. Such an approach improves responsiveness and resource allocation while decreasing the damages by strategically directing resources to the most crucial zones.

Intelligent defence architectures shape organisational practices and influence the allocation of resources and strategic decisions, shifting focus beyond traditional cybersecurity measures. Businesses adopting this approach can nurture a security culture and build resilience within their digital frameworks. Furthermore, the adoption of intelligent defence architectures demonstrates an organisation's efforts towards keeping up with technological changes, maintaining a strong security posture amid relentless cyberattacks and signalling proactive action on persistent security challenges.

Core Principles of AI Integration in Cybersecurity

Several fundamental concepts focusing on the function and importance of the integration of A.I into Cybersecurity serve as the foundation upon which it's built. Embracing artificial intelligence's capabilities when integrating A.I into Cybersecurity enables organisations to improve on security features, thus aiding in threat detection and neutralisation. Continuous learning and adaptation by AI vividly illustrate one core principle along with A.I. powered security devices which strengthen defences along with adapting to cyber threat evolution. Determined by adaptive approaches, these principles tend to work towards increasing efficiency, agility, and mitigating industry-specific vulnerabilities.

A.I helps reveal emerging threats by analysing complex datasets and assisting in uncovering patterns and anomalies. Automation enables other operators dealing with complex issues to concentrate on intricate factors of cyber defence due to emerging threats caused by unmonitored and repeatable tasks. This improves the operational effectiveness of the firm while increasing organisational response time in dealing with cyber incidents. Along with these principles, embracing A.I prioritises explainability and transparency, thus encouraging the importance of cyber defence frameworks that deal with comprehensible and scrutinised decision-making processes with figure accountability.

In the context of integrating AI within cybersecurity, ethical boundaries determine its use, reinforcing compliance and responsibility in wielding AI technologies. In general, all AI integration within cybersecurity strives to enable organisations with proactive, adaptable, and intelligent counter-offensive capabilities critical for enduring protections in the face of pervasive cyber threats in today's digital ecosystem.

Frameworks for Developing Adaptive Defence Mechanisms

The low adaptive mechanisms in the realm of cybersecurity have become a necessity. An adaptive defence is a proactive and dynamic method in relation to a cyber threat focused on actively monitoring potential breaches and responding to possible attacks. Focus will be placed on the fundamental frameworks for building adaptable defence mechanisms, describing the basic principles and concrete action plans. The overarching objective is to equip organisations with capabilities to counter prevailing emerging threats, leveraging real-time systems.

The key adaptive defence frameworks begin with a risk-based approach for decision-making. Organisations allocate resources based on the level of risk a particular threat poses. This entails understanding the threat landscape, identifying known risks, and reallocating resources to counter the most critical threats. Typically, organisations face challenges in accurately determining and allocating resources due to AI-driven predictive algorithms.

Integration of threat intelligence within adaptive frameworks is also equally important. As organisations focus on predictive analytics, they are capable of anticipating future attacks and strengthening their defences using gathered intelligence. Platforms empowered with machine learning are capable of scanning data, discovering hidden patterns, and forecasting relevant risks.

Resilience engineering is yet another advanced defence fundamental principle that influences adaptive frameworks. Organisations are capable of minimising the impact of incidents by fortifying systems with built-in resilience meant to withstand and swiftly recover from attacks. This paradigm shift moves from traditional strategies focused solely on prevention, adopting a more comprehensive strategy that emphasises prompt detection, potent containment, and swift recovery mechanisms.

Furthermore, the use of advanced analytics and behavioural modelling within the adaptive defence layers permits detection of anomalous activities and possible indicators of compromise. As organisations establish baselines of normal behaviour and use anomaly detection algorithms, they can respond to and mitigate suspicious activities, enhancing their security posture.

Adoption of a zero-trust security model also constitutes a fundamental pillar of adaptive defence. This model is based on the principle of 'never trust, always verify' and categorises all network traffic as untrusted until identification through rigorous authentication and permission controls is granted to users and devices.

Effective adaptive defence development requires integration of risk management, threat intelligence, resilience engineering, advanced analytics, and the zero-trust model. Organisational cyberspace defences can be strengthened and adaptive measures against constant changes in the threat landscape can be streamlined by applying these frameworks.

Harnessing Machine Learning for Threat Detection

The world of cybersecurity is facing escalating volumes of cyber threats as well as their sophistication, rendering traditional security methods obsolete. As businesses continue to grapple with advancing technologies, vicious activities, and incessant tactics, machine learning stands as a proficient companion in this technological war. The incorporation of machine learning in threat detection integrates its sophisticated algorithms that can analyse data over time and identify patterns that can foresee anomalies and breaches within the system. Thus, security systems can now proactively mitigate the consequences of such activities on crucial assets and infrastructure. With security offered through machines equipped for anomaly detection, pattern recognition, and behavioural analytics, organisations and businesses fortified with advanced technology and security will greatly enhance their capabilities to avert attacks that go undetected by conventional security methods. More so, machine learning empowers users with the ease of constructing predictive models based on routine historical data analysis, allowing adaptive modification and enhancement during the analysis of real-time data, hence assisting in preempting evil deeds.

The use of machine learning in enterprise threat detection systems enables an organisation to monitor and analyse enormous volumes of data such as network traffic patterns, user activities, and system logs to find masked traces of nefarious activities. Moreover, machine learning improves the efficacy and precision of threat detection by automating assessments of numerous data streams, as it is designed to enhance organisational efficiency, thus constricting the timeframe during which threats lurk within networks. Notwithstanding, the successful deployment of machine learning for threat detection requires profound knowledge of the organisation's security parameters, known vulnerabilities, data streams, and other situational determinants. It also requires frequent updates and adjustments to machine learning models to adapt to changing threat environments to lower the occurrence of false alarms and unrecognised real threats. Furthermore, the implementation of machine learning on an organisation's pre-existing security framework requires strong governance and protection of data privacy as well as ethical safeguards to control for compliance with laws when dealing with sensitive data. While organisations strive to balance the opportunities and challenges of using machine learning for threat detection, working with domain professionals, data experts, and ethical hackers becomes crucial in informing the design of thorough and robust counteractive frameworks.

Essentially, employing machine learning technologies in threat detection is transforming cybersecurity by enabling organisational preemptive action against advanced persistent digital threats.

Real-time Data Processing and Analysis

The continuous changes in the digital world have increased the demand for processing and analysing data in real time for purposes of cybersecurity. Timely and accurate evaluation of enormous amounts of data to detect possible threats and weaknesses is critical to effective defence strategies. Real-time data analysis is the actionable insight extraction from the stream by ingestion, processing, and analysing. Integrating these systems allows for security responses to be triggered automatically according to changed perceptions and to emerging threats and incidents. In addition to aiding responses to incidents, assessment of data in real time enhances understanding of the threats which require proactive countermeasures. With the use of Artificial Intelligence (AI) and machine learning technologies, it is possible for organisations to monitor large databases in real-time and pinpoint unusual activities or trends which may signify potential threats.

Beyond agility, real time powered analysis provides the responsiveness needed to adjust to multi-dimensional attack mechanisms and advanced cyber threats. Organisations are able to reduce chances of exposure to attacks by continuously monitoring and analysing data streams in order to manage threats more efficiently. Beyond this, real time analysis and prediction allows for the uncovering of more hidden relations that may not be seen using traditional methods of analysis.

This capability provides participating users with the possibilities of discovering intricate attack trajectories and forecasting probable future threats stemming from ongoing actions. Moreover, there are obstacles to be overcome with the implementation of real-time processes for data collection and analysis. Organisations have to resolve issues relating to privacy of data, system scalability, and the reliability of automated threat detection systems. Besides, the integration of real-time analytics is difficult due to the need for a strong architectural backbone and clear-cut organisational procedures that would allow for the quick computation and distribution of vital information. In any case, such data processing and analysis provides immense real-time benefits which enable organisations to adopt a proactive stance and agile response to perpetually changing cyber threats. With the progress of cybersecurity innovation, the importance of data processing and analysis in real time will never diminish when it comes to shielding against breaches and reducing risk.

Integrating Predictive Analytics in Defence Systems

The use of predictive analytics is an essential feature of modern defence systems and is fundamentally changing how organisations deal with cybersecurity. With predictive analytics, security teams can leverage both historic and current information to proactively identify and mitigate emerging risks and threats, thereby strengthening the institutions' defence pillars. Organisations can ascertain and even foresee potential threat indicators through complex algorithms and machine learning models that scrutinise and sift massive data sets to delineate the focal point of compromise, which may be lost in the traditional rule-based detection systems guiding most defences. Predictive analytics is equally important when it comes to improving the incident response capabilities of an organisation since the security teams are able to deal with problems while they are still developing. This approach reduces the losses caused by cyber incidents and at the same time improves the functionality of the defence systems.

One of the primary benefits offered by predictive analytics in defence systems is their ability to continuously recalibrate with respect to the evolving framework of threats. With intelligence data and historical attack data, predictive analytics can forecast likely attack vectors, equipping organisations with insights needed to proactively strengthen their defences. Moreover, predictive analytics assist organisations in identifying behaviour anomalies owing to changes which were normally invisible, sharpening the focus on activities that signal ultra-sophisticated cyber intrusions, even when those intrusions do not conform to established patterns. Taken in context, this stance is incredibly important when one considers that modern adversaries work around traditional security frameworks at every single opportunity to refine their tactics.

Adapting predictive analytics in defence systems goes beyond threat detection and encompasses strategic decision-making at various levels within the defence structures. Foreseeing emerging threats, possible attacks, and vulnerabilities allows leaders to plan, effectively train personnel, prioritise initiatives ahead of time, and allocate resources in a smarter way. This approach transforms the organisation's culture into one dominated by continuous improvement coupled with adaptive resilience while optimising the security framework of the organisation.

Still, there are challenges in the integration of predictive analytics in defence systems. Ensuring the reliability and accuracy of predictive models demands a rigorous approach to the quality of data, feature engineering, and tuning of algorithms. The governance and ethics on the use of predictive analytics, like bias in algorithmic outcomes and worries about privacy, require oversight of governance and robust ethical assessment.

With the evolution of the cybersecurity environment, the use of predictive analytics in defence systems remains an example of adaptive and proactive cybersecurity. With predictive analytics, an organisation can move beyond reactive security approaches to bolster their resilience and navigate the complex and rapidly evolving landscape of cyber warfare one step ahead of malicious actors.

Enhancing System Resilience through AI

As far as network security is concerned, system resilience is crucial. Cyber threats demanding up-to-date breach detection from systems are becoming more common, making the ability to withstand and recover from them highly advantageous. Proactive and adaptive defence frameworks are being implemented using advanced AI. AI enables quicker and more accurate anomaly and threat detection compared to older methods, allowing organisations to protect their systems before attacks can even be attempted. AI resilience goes beyond simply addressing identified problems and responding to them; it also includes risk assessment of emerging and novel threats.

One of the most important aspects of augmenting system resilience through AI is the appropriate technology framework for constant threat evaluation and adaptable solutions. Moreover, AI facilitates the automation of response plans, ensuring immediate and synchronised actions to security incidents. AI technologies are capable of swiftly containing breaches, mitigating both their scope and impact. Further, AI's capacity to learn from past attack strategies and corresponding countermeasures automates the refinement of defence systems, further reinforcing overall resilience.

Furthermore, the changes brought about by AI enable self-adaptive and self-learning systems to take shape, evolving in accordance with new threats as they emerge. These systems can autonomously identify new attack vectors and vulnerabilities, adapt enforcement mechanisms, and even anticipate macro-level risks, thereby strengthening organisational resilience against emerging vulnerabilities. In a world where cyber threats evolve in complexity and scale, the ability to accurately predict and adapt to emerging risks becomes a key strategic advantage, underscoring AI's pivotal role in enhancing system resilience.

Finally, AI allows organisations to move beyond passive responses to cybersecurity challenges and adopt a proactive and resilient posture towards those challenges. By strengthening system resilience with AI, organisations are better positioned to prevent potential disruptions, protect critical assets, and maintain operational continuity in the face of an evolving threatening environment.

Collaborative Defence Models and Information Sharing

In cybersecurity, attackers' techniques are becoming increasingly sophisticated, making breaches easier to perpetrate. Information sharing and collaborative defensive models are proving vital as businesses adapt to these threats. Enabling collaboration between peers from different industries, government bodies, and even vendors can help mobilise resources and intellect to enhance overall resilience. Collaborative defence models focus on proactive strategies where real-time responsive threat intelligence exchanges are made. Organisations are able to neutralise threats much earlier than they would have otherwise. Through the integration of active and passive collaboration, cross-domain expertise, and collective actionable insights, effective collaborative defence models enhance strategic understanding of the broad threat landscape, resulting in robust multi-layered defence strategies that ensure a stronger cybersecurity posture for organisations. From another perspective, sharing information also streamlines redundancy across businesses, resolves common problems, and consolidates duplicated work across multiple organisations, while optimising the use of scarce resources. These factors enable a stronger and more unified posture of collective defence against cyber threats. Conversely, the use of collaborative defence models may present privacy concerns, regulatory compliance issues, and a lack of trust between entities as significant hurdles.

Addressing these challenges involves creating effective governance models and standard operating procedures for safe information exchange. Additionally, the malicious exploitation of anonymisation techniques and the exposure of sensitive data is crucial in preserving the integrity of shared intelligence. Establishing a shared sense of trust and reciprocity is fundamental to incentivising participation in defensive collaboration. With the constant evolution of cyber threats, cross-boundary collaboration and information exchange become ever more critically valuable. Collaborative defence approaches should be adopted not only as a strategic option but also collectively as an obligation to protect cyber ecosystems. By sharing actionable intelligence and working together, organisations can actively strengthen their collective cyber defenses and be prepared to counter and eliminate cyber threats.

Challenges in Implementing AI-Driven Security Solutions

Deploying AI-driven security measures offers a multitude of challenges that require significant focus and planning to resolve strategically. One of the most critical issues is the lack of legacy systems integration with advanced AI technologies. Organisations are often stuck trying to integrate these new technologies with older systems, creating compatibility challenges and risking operational disruptions.

There is the rapidly evolving nature of cybersecurity which makes integration of different AIs and their platforms into one system a herculean task. Another critical challenge focuses on issues pertaining to AI ethics and operations within the security domain. As machines become capable of more critical AI tasks, the ethical dimensions of such decisions need to be assessed thoroughly. Finding a feasible solution where AI can be utilised to its optimum while dealing with ethical issues becomes the difficult part for security specialists. The lack of skilled individuals the industry faces compounds the issues even more. Retaining professionals with a dual expertise in AI and Cybersecurity has proven to be an everlasting challenge for organisations that seek to strengthen their strategies to mitigate cyber threats. Additionally, AIs dealing with security will need to change constantly due to the ever-changing nature of cyber threats.

The evolving landscape of threats and attacks requires AI technologies to be continually tuned and refined on their algorithmic and analytical capabilities. Uneven focus on emergent threats is likely to make ironic defences agile but unguarded, underscoring ongoing attention towards development. Moreover, making sure that AI technologies have adequate robustness and reliability in their algorithms when facing hostile adversarial forces is one of the most pivotal challenges in deploying AI-driven security innovations. Abstractions of such AI systems may be attacked and, in return, improperly disabled, crushing the efficiency of the security structure. Addressing such problems that need solutions while shielding algorithms from potential attacks is crucial to trust the touted enhancements provided by AI defences. Unsurprisingly, the challenge of compliance regulation alongside outlined legal contingencies layers even more complexity onto the scope shielded by AI-driven automation security. Tackling the perplexing maze of privacy laws, sectorial rules, and global law poses overwhelming difficulties for enterprises aiming to deploy AI technologies in their surrounding security infrastructures. Meeting impossible demands of ideas formulated, hoping to meet them while granting stricter mandates requiring AI limbs not to 'over' breach limits further complicates the ongoing process and begs to be monitored. Addressing these intermingling issues is very important in order to fulfil the promise of contemporary deep-seated issues of cyber security, heavily fortified by AI drawbacks.

Future Prospects and Evolution of Intelligent Defence Systems

The intelligent defence systems of intelligent cyber defence is a matter of priority discussion and most certainly speculation within the industry. Due to constant facing challenges from sophisticated inter-department cyber threats, advanced mechanisms of defence powered by artificial intelligence (AI) are crucial. The advancement of intelligent defence mechanisms will be dictated by several primary drivers concerning changes to cybersecurity in the next few years.

One of these key drivers is the adoption of AI with other emerging technologies such as quantum computing and IoT. This merging is bound to change the pace and scale at which the capabilities of defence systems will consider threats and respond to them. Moreover, the natural language interpretation and contextual understanding of AI promise a high level of interpreting complex and unstructured data.

The increasing focus on autonomous cybersecurity operations will also shape the development of intelligent defence systems. Greater autonomy in threat detection, evaluation, and neutralisation will strengthen defensive capabilities and lessen human intervention, thus minimising both response times and risks. On the downside, autonomy introduces unprecedented ethical and regulatory challenges that need to be solved in order to ensure that these systems can be safely used and operated.

The resultant paradigm shifts in cybersecurity will stem from the integration of intelligent defence systems with cloud structures and decentralised networks. This fusion will create the need for innovative ways to protect cloud-native applications and distributed computing environment's data, as well as address the inherent weaknesses of distributed computing environments. Security experts need to understand these changes and develop effective methods for protecting cloud-hosted resources from targeted assets in the face of emerging threats.

With the shift towards intelligent defence systems, traditional borders are set to be left behind as the need arises for unprecedented cyber defence. From the growing number of connected devices to the exponential increase of information being transferred across digital ecosystems, there now exists a need for fully advanced defensive systems which can provide security at every endpoint, network, and data repository. Besides, physical domains being infused with the digital world due to cyber-physical systems create new vulnerabilities and proactively protective measures need to be employed to defend critical infrastructures.

To summarise, intelligent systems of defence have the potential to reshape the paradigm of cybersecurity and trusting the evolution makes the intelligent systems full of promise. At the same time, we must ensure there is proper focus placed on legislative, ethical and social consequences on the nature of technology and society as a whole. Addressing the intelligent systems from the ground while respecting these regards will strengthen organisations in overcoming stubborn security threats penetrating the global landscape.

5
AI-Driven Threat Intelligence and Prediction

Introducting Threat Intelligence

As a major component of cybersecurity, threat intelligence is a constantly changing field. With the relentless advancement of cyber threats targeting businesses and individuals, developing comprehensive attack and incident response strategies that consider the intricate nature of cyber threats is vital. Threat intelligence, in simple terms, refers to the process of gathering, analysing, and interpreting information aimed at predicting possible cyber threats and cyber vulnerabilities an organisation may face. Organisations can obtain valuable insights on malicious intent, attack trends, and risks with the help of internal and external data sources. This helps provide context about the threats businesses face, allowing them to thwart potential threats before they escalate into damaging cyber incidents. Furthermore, threat intelligence helps security teams appropriately allocate and prioritise their resources, focusing on the most vulnerable areas. This, in turn, allows the organisation to bolster potential weaknesses that could be exploited by threats. Building proactive defence mechanisms requires a strong threat intelligence backbone. This is critical for organisations to deal with advanced persistent threats as it provides the understanding of threat actors' tactics, techniques, and procedures, thus reducing the element of surprise and disruption posed by cyber attacks.

Threat intelligence sharpens incident response capabilities so that an organisation can decisively act on security breaches within moments of detection, thereby limiting the damage to the organisation and mitigating further exposure. In the context of a digital environment where interconnectedness is the norm, not having threat intelligence is akin to a business omitting an essential function in their operating processes; thus, an organisation cannot effectively protect its assets and operational stamina without threat intelligence. Effective threat intelligence helps businesses make appropriate decisions and establish security measures in accordance with the company's risk profile and business goals, which, in turn, enables optimal business outcomes.

Data Collection and Analysis Techniques

Within the domain of cybersecurity, the techniques for data collection and analysis form the foundation of threat intelligence. The process begins with collecting data from multiple sources like network logs, endpoint telemetry, threat intel feeds, and even open-source intelligence. After collection, this data goes through normalisation, categorisation, and correlation to unveil actionable insights to refine data. In the context of cybersecurity, data mining, pattern recognition, and anomaly detection are necessary to sift through enormous datasets to identify potential threats and vulnerabilities. Besides, the analysis of unstructured data, such as social media posts and forum discussions, by integrating natural language processing (NLP) can help detect trends or emerging threats in advance.

The analysis phase includes interpreting the filtered data with the help of statistics, machine learning, and behavioural analytics. As explained earlier, statistical techniques provide quantitative insights with respect to historical attacks and their patterns which is also crucial to establish baseline behaviours for anomaly detection. Attack patterns are analysed using machine learning models, both supervised and unsupervised, to detect known and unknown threats by invoking recognised distinct patterns or deviations from established normal activity. Lastly, behavioural analytics track changes in activity that differ from accepted norms or expectations and classify them as anomalies to expose malicious activities that are otherwise undetected by signature-based systems.

The integration of artificial intelligence technologies with threat data analysis improves the efficiency and precision of identifying threats. Organisations can forecast and estimate a probable course of action of the attackers several steps ahead, and scale their response actions in the case where preemptive actions are warranted. For instance, in clustering, a number of algorithms exist that take incidents as their inputs and group them into classes based on similarity for the purpose of uncovering higher-level common attack patterns that can be used for more effective decision making on countermeasure planning. Artificial intelligence enhances the analysis step of the data processing cycle by making connections of events that may seem unrelated to humans but reveal complex sets of attacks and orchestrated campaigns far beyond human capabilities.

In the ever-evolving cyber environment, the methods of data collection and analysis require constant refinement and validation. Updating threat models and machine learning algorithms regularly maintains versatility, while automated analysis with feedback from human analysts improves responsiveness and accuracy in identifying false security breaches and genuine breaches. Ultimately, advanced persistent adaptive cyberspace attacks on an organisation's critical assets require precision and agility in data collection and analysis, sophisticated techniques for evolving threats, and fortifying systems to ensure operational continuity.

Predictive Analytics in Cybersecurity

Within the cybersecurity landscape, the implementation and effectiveness of machine learning models are invaluable for the detection and mitigation of security threats. These models use sophisticated algorithms to analyse extensive datasets, enabling security teams to find suspicious patterns and anomalies. Normal and malicious behaviours are differentiated with supervised learning techniques and labelled datasets. Moreover, unsupervised learning approaches are capable of identifying previously undetected threats by identifying deviations from expected behaviour.

One of the most popular machine learning models for anomaly detection is the anomaly detection model, which aims to find outliers in datasets. Through training on historical data with current network traffic, this model can quickly detect and alert security personnel to activities that significantly differ from expected norms. Furthermore, decision tree models effectively classify and rank potential threats on diverse parameters, which assists security analysts in focusing their responses.

An important implementation of machine learning in threat detection is the application of neural networks. Neural networks are sophisticated models derived from and meant to emulate human cognitive processes. They analyse intricate and multidimensional data to identify profound patterns. Thus, they are very effective at spotting potential cyber attacks far in advance, even when such attempts are meticulously disguised to bypass traditional rule-based systems. Moreover, neural network technologies enable cybersecurity professionals to explore non-linear interrelationships within datasets, which helps strengthen holistic security systems.

Also, ensemble learning is another model used in threat detection. It focuses on increasing prediction accuracy by combining multiple models. Ensemble models utilise distinct algorithms and amalgamate the outputs. This approach is extremely useful in dealing with evolving threats because it offers considerably stronger protection. Because of their unique capabilities to address some of the critical issues in cybersecurity, ensemble models are important in protecting organisations from sophisticated access and attacks.

Nonetheless, an equally important consideration is that machine learning models, which can be extremely useful for threat detection, have certain shortcomings. Specifically, adversarial attacks, where malicious individuals attempt to input deceptive data to confuse models, is a major issue. Furthermore, the cyber threats themselves require a lot of focus, so these machine learning models need to be constantly revised and recalibrated. In sum, machine learning models offer a very active and vital approach to contemporary cybersecurity functions, automating threat detection.

Integrating AI with Existing Security Frameworks

With the growing sophistication of threats, cybersecurity experts are now using predictive analytics to foresee and mitigate risks. Organisations are better prepared against cyber threats by using historical and real-time data to predict imminent risks. In regard to cybersecurity, predictive analytics encompasses all machine learning algorithms, statistical models, and behavioural analysis. These techniques make it possible for security teams to detect patterns and anomalies that could signal breach attempts days or weeks in advance. By utilising available data such as network traffic, user behaviour, and other system logs, security practitioners can avert major disasters and proactively patch vulnerabilities before they are exploited. Aside from forecasting breaches, predictive analytics can also assist in forecasting potential attack vectors to help design preemptive countermeasures against targeted worrisome threats.

The utilisation of predictive analytics within cybersecurity allows organisations to avert and mitigate debilitating cyber threats before they occur. The systems' anomalous and pattern recognition behaviour can alert organisations on system activities that could turn out to be breaches if left unchecked. Moreover, anomalies that go undetected or ignored also fall into these patterns. With the aid of predictive analytics, security teams are also able to prioritise alerts in a critical manner based on their severity. This streamlines not only the response to the alert but also the allocation of resources. The operational efficiency is improved by this precision focus approach along with enhancing reliant response systems.

With the integration of external threat data like indicators of compromise (IoC) or new attack trends, models that utilise external indicators can implement advanced analytics to estimate and predict information pertaining to internal networks. Thus, internal network activity with external network information provides a thorough understanding of dangers alongside a proactive defence approach. Predictive analytics' capability of tightening organisational cyber defenses proactively greatly enhances cyber threat intelligence.

As has been emphasised already, predictive analytics continues to stand out as one of the most impactful weapons in the field of modern cybersecurity. As a result of its unique capability to transform unstructured information into meaningful insights, it enables an organisation to also reinforce their cyber defences while biologically adapting two steps faster than their cyber foes. As explained previously, predictive analytics actively increases intelligence on determined threats, proactively addresses predictable gaps, fortifies the cybersecurity perimeters of invaluable assets, all while wiping out predictable security incidents long before it can occur.

Real-Time Monitoring and Alert Systems

Amid unprecedented assaults of evolving cyber threats on any business infrastructure, augmenting AI with the pre-existing cyber security protocols is now necessitated as the primary means of activating resiliency enforcement measures. With AI's high-level analytical and anticipating capabilities, it optimises existing reactive approaches by enabling security personnel to proactively pinpoint, scrutinise, and counter potential breaches with more effectiveness than pre-existing methods provided. Having AI integrated with current cyber security frameworks enables organisations to improve their threat detection through the effective application of machine learning algorithms empowered to quickly analyse massive chunks of data for malicious patterns.

AI improves formerly static security systems with real-time adaptive and dynamic response mechanisms. By analysing the shifting threat environment, AI technologies can independently fine-tune security protocols for active threat shielding. This flexibility is essential today, when the frequency and sophistication of cyber assaults are grappling with unprecedented levels. Security frameworks powered with AI provide advanced visibility and probing capabilities over network activity streams, which allows security personnel to identify suspicious activities and potential breaches faster and more accurately than before.

Another important aspect to recall is the use of AI within existing security frameworks and its integration within the processes of responding to incidents, in particular how AI-driven insights can provocatively alter workflows with no manual adjustments. When anomalies or threats are flagged, response workflows can be automated to deal directly with the incident at hand by issuing alerts and automating containment procedures that are executed instantly at different levels of escalation. Such automation streamlines complex workflows which means organisations will operate on near real-time, thus reducing impact from security incidents and slashing losses.

Regardless, the combination of AI technologies and pre-existing security framework systems reveals challenges. Ensuring the application of AI technologies' interoperability and compatibility within diverse security infrastructures involves resourceful design and implementation. Additionally, ensuring data privacy, algorithmic bias, and AI model ethics and other issues is equally significant in sustaining the trust and transparency expected in security operations with AI technologies. Thus, there is a need for elements of comprehensive security planning by merging the use of AI technologies alongside human intelligence in order to protect an organisation's digital assets.

To summarise, the use of AI technologies alongside existing security frameworks can augment organisational defenses against contemporary cyber threats. The adoption of AI technologies to improve threat detection, mitigate risks, enhance real-time adjustment, or automate incident response enables an organisation to assume control in actively managing risks and protecting critical organisational assets. However, comprehensive integration calls for a balance of technological, operational, and ethical factors, requiring synergy between AI systems and domain professional practitioners.

Case Studies: Successful Deployments

In the ever-changing world of cyber threats, real-time monitoring and alert systems offer proactive measures that provide a first line of defence. System alerts and intelligence notification utilise AI algorithms to monitor network traffic, system logs, and applications on a 24/7 basis for any anomalies. Organisations are enabled by AI pattern recognition and anomaly detection algorithms to easily uncover breaches of security parameters or attempts to trespass into their networks.

As a feature let alone a system on its own, real-time monitoring focuses on the ongoing stream of data that is continuously flowing in. AI tools placed to monitor a system strengthen an organisation's defence by identifying potential threats marked out by lower levels of false positives. Such systems also update themselves based on new attack techniques, thus placing organisations ahead of attackers trying new approaches.

The integration of predictive analytics to real-time monitoring systems strives to go beyond alerting to threats by actively seeking to defend against them. Strong surveillance systems assist security teams to effectively and efficiently preempt target anticipated flights or easily branch to accessible routes. With the ability to analyse historical data and trends, AI systems are empowered by such predictors which allow organisations to proactively resolve risks and issues before they happen.

The effectiveness of contemporary systems for monitoring security issues in real time is augmented by the contextual information they provide on specific threats. Threat detection and user behaviour analytics in a network can be sufficiently advanced to integrate disparate information and offer security analysts holistic supplements regarding possible risks. Strategic prioritisation of alerts and efficient incident response within a particular timeframe enhances detection and shortens the chain for remediation of security incidents.

Automated actions taken as a result of continual monitoring can greatly benefit firms or companies. Through AI-powered orchestration and response protocols, some predefined steps can be taken in absolute automation. For example, actions such as containing compromised systems or blocking active malicious actions; and thus reduce the impact of security breaches while lowering the level of human control required.

All in all, the use of monitoring and alerting systems with embedded AI technology appears to be a core element of a modern cyber security approach. Continuous monitoring and proactively scouting fears using AI underpins broad defence mechanisms to avert threats faced by firms and other organisations.

Challenges and Limitations of AI-Driven Methods

Here, we focus on the successful real-world applications of AI-powered threat intelligence and forecasting systems. Theory is always better understood when it is applied to practical scenarios. Our first example involves a financial institution that enhanced their cybersecurity frameworks using threat AI. The institution's machine learning systems were capable of identifying new threats and predicting several prospective attack vectors with unprecedented accuracy. This enabled the organisation to neutralise advanced cyber attack attempts on the institution, preserving customer trust and sensitive data. Equally compelling is the example of a multinational technology company that used AI based threat information and fortified their network perimeter security against advancing cyber threats. Through constant scrutiny of network data traffic, the company achieved critical visibility and preemptive threat mitigation AI aided detection. They were able to prevent costly breaches and system downtimes by using AI and their existing infrastructure. To add to this, we shall outline another example, where a healthcare institution used tools designed to predict and provide AI-based threat intelligence for the successful protection of their systems and data.

Through predictive analytics, the organisation was

capable of remediating high-risk vulnerabilities within their network, fortifying critical cyber attack pathways long before threats could emerge. In doing so, they safeguarded imperative patient records, sidestepping stringent penalties for breaches of data privacy regulations. Studying these cases illustrates the significant impacts that solutions powered by AI have on organisational adaptability as well as the agility to respond to modern cyber threats. Predictive intelligence and AI-driven threat frameworks became pivotal in fortifying digital assets and vital infrastructures from safeguarding heinously sophisticated malign attacks.

The Role of Human Expertise in AI Systems

Every emerging potential of the preventive measures fuelled by AI is burdened with its own challenges and imperfections that seek acknowledgment before being utilised as a solution in systems this sensitive. Consternation arises from ever-challenging and diversifying world of cyber-attacks; The additional revolutions an attacker brings with themselves push the boundaries of what can be exploited beyond unimaginable lengths. The moment an AI is granted control with the intention of defending against such loaded thrusts, it loses the ability to respond to the dynamic nature of our real world, swiftly rendering it useless. Risk of losing focus on the detection and response to attack grows especially when dealing with new unseen situations due to solely training AI models on past data.

Another important limitation relates to the lack of interpretability and explainability of methods based on Artificial Intelligence (AI). The systems in question can effectively recognise patterns, clusters, and outliers in enormous datasets, yet the rationale behind their decisions presents a considerable challenge. The issue of algorithmic transparency brings forth questions of responsibility and trust given that security practitioners would find it difficult to comprehend and verify the logic behind AI insights and recommendations.

AI methods are vulnerable to adversarial attacks aimed at deliberately manipulating system inputs to alter the functionality of an AI system. This vulnerability poses a fundamental concern regarding the integrity and reliability of AI-based cybersecurity solutions.

Furthermore, the lack of an adequate number of AI-driven cybersecurity tools professionals creates new problems. Most organisations are unable to attract or retain personnel with sufficient skills in cybersecurity and AI, which slows down the adoption of these sophisticated technologies.

Moreover, AI techniques for threat detection and prediction carry responsibilities of privacy, ethics, and accountability against changing data protection laws and regulations. Thus, such methods require meticulous compliance strategies reinforcing fairness, responsiveness, and responsibility.

Future Trends in Threat Intelligence

While these challenges and constraints must be considered, it is important to understand the advantages these new AI methods offer and explore the vast topics relating to cybersecurity processes. By overcoming these problems through continuous research and collective efforts, innovation will emerge and technological advancement in these fields can be accelerated with lowered risks.

Significance of Human Intelligence In AI Solutions

As informatics grows steadily and computer systems become more common, human influence is critical in attention modulation and instruction for streamlining and improving AI systems' functions. AI technologies have a great role to play. Human beings still physically review important files and documents for information patterns that might emerge during trend analysis.

In relation to threat identification and assessment, harnessing human skills is essential for several reasons.

First and foremost, human analysts have a deep understanding of the context and knowledge relevant to the field with the capability to comprehend intricate systems which involve relationships and are AI algorithms. With this knowledge, they can confirm insights given to them by AI, disregard irrelevant details that are flagged as important, and understand the broader context of emerging threats. Human analysts are also useful in setting the boundaries of parameters which the AI models will use to ensure that algorithms designed fit the organisation's specific operational needs and threat landscape.

Evaluating the ethical aspects stemming from AI decision-making in cybersecurity relies heavily on human judgment. While any intricacies involving ethics such as privacy filters, compliance with the law, and other social implications are the work of humans, detection of fraud systems and other devices is no doubt an area that is covered by AI. An AI expert can study issues dealing with the different areas of ethics, especially AI and give advice on what to do in order to ensure systems that meet ethical and regulatory requirements are adopted.

Additionally, humans are essential for the ongoing improvement and refinement of AI systems. Through the continuous feedback of human analysts, who play an important part in the validation, additional attacks evolving AI models are adaptive feedback prediction of to Tech tech threats are validated, threats, and a new model of attack confirms and validates adapting attack tactic strategies while iteratively adapting to evolving attack tactics and strategies. This approach builds a symbiotic AI and human system designed to strengthen and enhance adaptive cybersecurity protection.

As organisations depend more heavily on AI systems for threat intelligence and forecasting, the need to incorporate and recognise human skill becomes critical in order to develop more holistic and effective cybersecurity defenses. There is a need to integrate the insights provided through technologies in AI with the nuanced, ethical human considerations in order to enable the organisation to anticipate and timely respond to emerging cyber challenges.

Upcoming Developments in Threat Intelligence

The use of advanced technology in threat intelligence will continue to evolve with the integration of artificial intelligence and machine learning. Emerging technologies are expected to influence the threat intelligence ecosystem in the following primary ways. The first is increased adoption of AI-powered cybersecurity solutions that will enable the design of more complex adaptive threat detection and response systems. This AI-Cybersecurity convergence will provide adaptive organisational resilience through predictive, prescriptive and real-time automated counteractions on prevailing and emerging threats. Furthermore, the paradigm of big data and cloud computing will transform the collection, processing and application of threat intelligence. With big data technologies, security analysts can process massive amounts of data in real time and detect patterns or anomalies that suggest possible security breaches, resulting in advanced threat detection. Additionally, the integration of threat intelligence platforms with AI-powered predictive analytics will allow organisations to foresee and prevent security risks, effectively countering advanced persistent threats long before they evolve into fully-fledged cyber-attacks. The other one to notice is the development of threat intelligence ecosystems where organisations participate and share threat information to advance collaborative counter/defensive mechanisms for network protection.

This cooperative method not only fosters a deeper understanding of developing risks, but it also enables the swift sharing of actionable intelligence throughout the cybersecurity ecosystem. As the scope of the Internet of Things (IoT) expands, the concept of threat intelligence will evolve to target a wider array of associated devices, which will give rise to new possible points of exploitation. IoT interconnectivity will require security professionals to rethink their threat intelligence methodologies and address the peculiarities of interconnected device challenges. Moreover, the emergence of adversarial AI will require the immediate formulation of AI-designed counteroffensives that target such attack vectors. There is no doubt that as attackers take advantage of highly advanced AI technologies to structure infiltration plans, an equally advanced AI counteraction will be demanded from the side of the defenders. Lastly, organisations will realign their security postures with ever-shifting legislative demands due to the constant change of compliance requirements and legal frameworks, fueling the evolution of threat intelligence. Intelligence and defensive capabilities will have to be refined to respond to evolving laws and regulations on data protection and privacy while ensuring strong, proactive positioning.

Ultimately, advancing technologies and collaborative approaches offer organisations fresh possibilities to protect their digital assets in cyberspace. This, however, comes with a continuously evolving threat landscape.

6

Automated Incident Response & Recovery

Overview of Incident Response Automation

As organisations face an ever-growing number of cyber threats, incident response is becoming a critical area of focus in the field of automation. In this case, we want to look at the advantages offset by automation in regards to incidents matters and how it affects the use of AI tools as well as machine learning as these tools increase the workload of security teams in eliminating and mitigating incidents. With incidents response automation, there is the possibility of significantly improving the time used in detecting, analysing, and responding to the security events. This also minimises the impact of cyber-attacks. Automation, on the other hand, improves the efficiency of scaling effort that the security teams are incurring as they are able to handle increased unlimited volume of incidents irrespective of resource levels. Automated response also offers capability to ensure accuracy and consistency of actions as there is reduced human errors from SLA breaches due to pressures associated with minutes remaining on the clock. Businesses can also employ the AI models for detection and alert escalation whereby alerts will be automatically prioritised based on severity levels and impacts. This gives the organisation better and faster identification of threats and enables centralisation to focus attention on the most demanding risks.

Moreover, the use of AI-driven playbooks and workflows offers automated incident response solutions, enabling rapid response containment strategies. Designed with predefined steps for execution, these playbooks contain threats and mitigate damage to prevent further spread. Automated root cause analysis ascertains the automated processes during incident responses addressing the primary origin of security breaches which enhances the system's resilience and security posture. Timely response neutralisation, proactive defence, and other machine learning techniques also bolster incident response efficacy by adapting to patterns and evolving threats. Additionally, recovery automation and restoration techniques after a security incident significantly reduce operational downtime and accelerate the return to business as usual. Automated incident response tools may operate in tandem with other cybersecurity solutions seamlessly through integration with the existing security frameworks, thereby improving the organisation's security posture. While dealing with sophisticated modern cyber threats, organisations need to evaluate automated responses using metrics and key performance indicators in order to fine-tune and optimise strategies. Along with the advantages discussed above, I would be remiss not to talk about the challenges and limitations such automation, internal opposition to change, encouraging compliance to external regulations, and the more commonly cited, false positives and negatives.

In the future, trends predictive of incident response automation seem to be sharper AI development, better synergy and collaboration of security tools, and merging automation with AI and human skills for multi-layered defence.

AI Models for Detection and Alert Prioritisation

Alert fatigue is a common phenomenon in cybersecurity incidents because of the frequency, volume, and intricacy of security alerts. This phenomenon can overwhelm analysts with a deluge of security alerts, potentially leading to delays in incident response. This challenge arises the need for effective detection and alert prioritisation systems utilising advanced AI PyTorch models. AI-based detection uses machine learning algorithms coupled with previous data sets' baseline behaviours to identify patterns and repeating occurrences, trends, as well as abnormalities. These AI models provide maximum detection accuracy in cyber-attacks by learning current data patterns and analysing present data alongside previously recorded data patterns. Moreover, AI systems can hone cybersecurity responses by anticipating the alert that requires attention first. Therefore, optimising resources with automated triage systems that analyse risk levels and prioritise pertinent incidents enables organisations to improve cyber attack responses. Thus, the firm's position about security resilience improves alongside response times and resource allocation. The main aspect of alert prioritisation in AI systems is the focused adaptability, self-correcting optimisation, and learning feedback loops.

These models evolve with constant feedback cycles and reinforcement learning to identify new patterns of concern and improve their threat detection algorithms. This ability helps organisations counter adaptive attacks and mitigate risks. AI models make it possible to integrate disparate data, such as network traffic, endpoint logs, threat intelligence feeds, and user behaviour analytics, giving a full picture of possible security incidents. With this good coverage, AI boosts the alert prioritisation accuracy and precision, allowing organisations to deal with threats proactively. As organisations adopt AI for detection and alert prioritisation, it becomes crucial to build strong governance and oversight mechanisms regarding these frameworks. AI enhances human work, and thus there should be clear mechanisms in place to guarantee that the processes carried out using these models retain sufficient transparency and accountability. Besides, the application of AI models should obey the established legal framework and ethical guidelines, pointing to the need for the responsible use of AI in cybersecurity. To conclude, the use of AI in detection and alert prioritisation constitutes a major advancement in enabling organisations to respond to incidents with heightened agility and efficiency.

Rapid Containment Strategies Using AI

For any organisation, averting damages while incident response during cyber threats is critical in today's world. Embracing new technologies, particularly artificial intelligence, rapid incident response strategies have become attainable. In this piece, we discuss the methodology of AI application in swift security incident containment.

One of the primary systems and networks under the organisation's purview, enabling instant anomaly, threat detection, and hazard recognition. With the escalation of cyber crimes, modern organisations require monitoring systems that can efficiently track and evaluate network operations such as traffic, system logs, and user behaviour in real time. AI systems are capable of responding adequately countervailing to assaults on multiple levels using the latest technological trends in machine learning algorithms. This helps such systems to autonomously respond to harmful activities, aiding rapid fracture containment.

The automation of response actions is another important facet for rapid containment using AI systems. Due to automation decision making and response playbooks created ahead of time, organisations can fully automate containment actions such as within system isolation, stoppage of hostile interactions, and containment of infected assets. This method of automation enhances both the speed at which containment is achieved and the time manual efforts take, therefore improving the possibility of consistent results no matter the differing incident scenarios.

AI can also perform predictive containment based on previous threats and behavioural analysis to determine proactively attack paths towards critical assets. Through AI capabilities incorporating various data sets and spotting early signs of identified threats, automatic systems can employ preemptive containment procedures aimed at strengthening the organisation's defences before an incident occurs. The overall approach fortifies the organisation's defences before an incident occurs, thus enhancing the security posture and reducing the damage caused by a security breach.

It is crucial to note the application of AI in strengthening human capabilities during rapid response and containment processes. Although AI tech specialises in large-scale info processing and analytics, human intervention is crucial in validating actionable items from AI insights, adapting evolving response plans to new dangers, and refining holistic efforts. Through providing actionable intelligence and insights through AI, organisations can empower cybersecurity professionals to foster collaboration between human judgment and AI automation through reliable and agile containment actions.

The integration of rapid response strategies and AI offers a game-changing development in the cybersecurity arms race. Advanced persistent threats require outsmarting and defeating with AI-driven rapid detection systems, automated response orchestration and predictive containment, powerful defences are made when harnessing human and machine intelligence. If organisations expect to stay ahead of emerging threats, the adoption of AI-driven rapid response and containment systems is non-optional for companies looking to enhance their adaptive efficiency to tough security situations.

Automated Root Cause Analysis

Within the context of cybersecurity, automated root cause analysis forms one of the phases of AI-enhanced incident response and recovery workflows. It involves the resolving of issues that manifest in the form of security and cyber incidents, systems downtimes, and performance lags. Automated root cause analysis seeks to optimise the identification and resolution processes of complex security events by exposing their primary triggers through advanced algorithms and machine learning techniques. In achieving this, organisations are able to enhance their incident response and management workflows while mitigating the adverse effects of cyber threats on business operations.

Automated root cause analysis implementation consists of critical stages. Firstly, there is collection and analysis of data pertaining to security incidents, network activities, and a system's infrastructure in real time. Through the application of pattern recognition algorithms, correlation analysis, as well as anomaly detection, autonomous hypotheses for the breaches of security and operational lags are actively generated. After that, the system can suggest possible solutions to the above insights and determine the most important factors that need to be addressed in order to initiate rapid remedial action.

The automated processes for analysing the root cause focus sharply towards reporting accurate results with the greatest efficiency as directed by machine learning. Patterns relating to security breaches are accurately mapped out by AI algorithms by analysing historical data as well as data streams from active monitoring of ongoing threat landscapes. The system now has the capability to identify behavioural anomalies, recognise new attack angles, and anticipate potential security loopholes. Organisations are able to mitigate risks and bolster overall resilience by eliminating foundational weaknesses that give rise to major security incidents long before they are triggered.

Integrating automated root-cause analysis with incident response orchestration and the recovery mechanisms provides significant results. Organisations can automatically execute containment, threat elimination, and system restoration by integrating root cause analysis with response playbooks and recovery procedures to which remedial actions have already been programmed. This integration helps in achieving optimal response times, automation, reduction in the number of people necessary for a task, and uniform application of critical best practices across various security situations.

Any automated root cause analysis will still face challenges. Modern IT frameworks, overwhelming volumes of event data, and ever-changing cyber threats severely hinder precise actionable insights. Furthermore, the non-explainable nature of AI models used for root cause analysis raises issues of accuracy and AI reliability that need constant tweaks to maintain trust.

Looking forward, integrating explainable AI, incremental graph analytics, and self-explaining models will enable automated root cause analysis to develop further. Predictive analytics will offer a new light to the root cause analysis architecture as well further expansion of autonomous engagement frameworks will merge with other AI-empowered capabilities turning it into a more multifaceted and responsive framework.

Machine Learning in Threat Mitigation

Organisations are approaching threat mitigation in cybersecurity through the lens of machine learning. The sensitivity and nature of security threats require advanced algorithms and statistical models, more so than any cyber system can utilise, to respond, detect, and analyse the issues. Machine learning systems continue to evolve, improving their ability to assist in modern cyber defence. They are pivotal in sifting through mountains of data to identify patterns and anomalies which can be used to mitigate security challenges preemptively.

Alerts can easily be differentiated through the classification and prioritisation features modern machine learning offers. Traditional rule-based systems face challenges distinguishing noise versus alerts. Machine learning systems, however, learn from historical data and dynamically adapt alerting mechanisms offering better context-aware alerts in real-time.

Further still, machine learning permits predictive threat intelligence, evaluating real-time data against a timeline of events. Observations from the past and present can be woven together to reveal potential threats that have not yet been actualised. With appropriately tailored algorithms, machine learning can identify ominous subtle signals of attack that most systems will miss. This underscores the importance of minimising the chances of, and consequences of, cyber incidents.

Beyond detection and prediction, machine learning facilitates responsive and proactive decision-making in real-time as new threats emerge. Anomaly and behaviour monitoring involve the application of machine learning algorithms to system activities to detect anomalies and take appropriate control measures automatically. This adaptive method of mitigation guarantees that organisations receive and respond to emerging threats in real time, thereby reducing the opportunity for exposure and damage.

As much as machine learning can be applied to mitigate threats, it will not be very effective if the training data is of low quality. For machine learning systems to accurately identify malicious activities, robust and diverse datasets need to be built so they can differentiate between positive and negative activities. Unceasing assessment and adjustments of the models are needed to maintain the efficacy of the machine learning frameworks in combating evolving cyber threats.

Organisations will continue to be faced by sophisticated cyber adversaries, thus the need to equip cybersecurity operations with machine learning tools will now and in the future, be invaluable. The combination of a human operator and machine learning will help redefine the future of threat mitigation and allow for better movement by the cybersecurity teams in the increasingly complex and volatile digital world.

Recovery Automation and Restoration Techniques

Prompt and effective recovery from a cyber incident is crucial in reducing an organisation's operational and reputational damage. In the case of AI-enabled cybersecurity, automation of the recovery processes is particularly important because it expedites the response, thus lessening the vulnerability window. It is important to explain the complex domain of recovery automation and highlights sophisticated restoration techniques.

One of the most important elements of recovery automation is the rapid detection and containment of infected systems or data. AI mechanisms can automate the rapid triage of affected assets, which allows the security teams to better allocate their efforts and assets. AI can evaluate compromise indicators (IoC) and behavioural anomalies in real-time to assess breach impact and automate infection zone containment with minimal human guidance.

Furthermore, the automation framework integrates the orchestration of specified restoration actions. AI systems can suggest and implement specific recovery workflows for particular scenarios using machine learning and historical incident databases. These workflows may include, but are not limited to, restoring to specified system snapshots, validating and deploying patch updates, or resetting access control configurations for certain networks. AI automation implements these recovery actions with unparalleled speed and precision.

Beyond the considerations of the technical recovery

procedures, AI-driven systems contribute to business restoration by automating overseeing actions for business continuity planning. AI systems automate plan activation for trimming critical business functions and performed services interdependencies, which minimises disruption to core functions. This recovery approach accelerates full technical infrastructure recovery and reduces functional impact on the organisation.

Furthermore, predictive analytics applied to recovery automation enables organisations to anticipate threat actors and dynamically adjust preparedness posture in advance. Using historical incidents datasets and new threat intelligence, AI models can analyse and identify system weaknesses which can be reinforced by automated security posture adjustments aimed at thwarting preemptive counteraction by adversaries. In this way, recovery automation is no longer a reactive procedure but instead acts as proactive defence enhancing the organisation against continuously recurring cyber-attacks.

To ensure maximum effectiveness, recovery automation must be aligned with incident response procedures and the peripheral security architecture. The automated recovery processes are purposefully avoided when there is non-compliance with higher level organisational security policies, as well as regulatory and compliance obligations, guarantees procedural compliance. Thus, holistic organisational cyber incident management systems can confidently and swiftly adapt to dynamic threat environments.

Integration with Existing Security Frameworks

In the quest to enhance an organisation's cybersecurity posture, incorporating automated incident response and recovery techniques with the existing security frameworks remains a significant concern. Integrating AR techniques with traditional security integrated systems requires planning that aligns with the organisation's security architecture. Our focus here is on the intricacies of incorporating automated incident response and recovery systems with existing frameworks, illuminating guiding principles and misconceptions.

An important concern in the approach is the deficiency of interaction between automated incident response systems and other existing security technologies. For effective holistic threat management, interaction with other supporting automation systems such as SIEM, firewalls, and intrusion detection systems is critical for seamless automated system communication, data handling, and exchange. As the appendium part of the existing security systems framework, the automated system can be integrated to improve system-wide agility and adaptive capacity to respond to threats.

It is also important to align the processes of automated incident response and recovery with organisational security policies and compliance requirements. All mapped processes need to define how the automated mechanism will respect relevant policy and security frameworks while executing incident response steps. This alignment will not only facilitate compliance with regulations but also provide a singular posture in resolving security incidents, thereby reducing friction to business processes.

Another important area that has to be looked at concerning the automated system is their responsiveness to different technological settings. In regards to an organisation's presence on premise, in the cloud, or hybrid, the automated incident response and recovery system should be able to function efficiently in all environments without disrupting operational effectiveness. Such uniformity may require adjusting automation workflows and responses to specific environment particularities so that comprehensive security operations are achieved.

Additionally, integration with other security components should have strong response orchestration and coordination features. This includes setting specific escalation routes, assigning and tracking given roles, and aligning the automated incident response to human actions when necessary. Striking a balance between automated and manual response frameworks enhances collaboration towards the organisation's defence which, as a result, strengthens security incident response and containment capability.

Finally, ongoing evaluation and enhancement are critical to the integration process. Constant observation of the performance of the integrated system within its operational environment, along with proactive adjustments, is essential. This iterative process enables organisations to strategically navigate and automate incident response and recovery processes when applicable within their security infrastructures.

Evaluating the Success of Automated Responses

In today's world of cybersecurity, initiatives such as automated incident response and recovery are vital for the effective handling of security events. Automated responses are evaluated to check their reliability and effectiveness in mitigating threats, which requires comprehensive analyses and investigations of a series of critical processes (KPIs) and metrics to evaluate the effectiveness of employed automated incident responses.

A primary form of evaluating automated responses is measuring the change in mean time to detect (MTTD) and mean time to respond (MTTR). These metrics show the responsiveness of an automated system concerning detecting and solving security incidents. Decrease in both MTTD and MTTR metrics signifies a beneficial effect from the automation in addressing security incidents.

Along with participant evaluation, AI-driven contextual responses require accuracy and precision as critical metrics for assessment. It is equally important to evaluate the automation's metrics on false positives and false negatives. A low false positive rate means the automation is accurate and effective in identifying actual security threats, which reduces alert fatigue. Moreover, a low false negative rate means the automation assesses real security incidents accurately and responds promptly, minimising the chances of organisational risk oversights.

Determining response adaptability and scalability are critical in evaluating automated reactions. In unmanaged systems, agility is a core facet for the automated response mechanisms to adapt to their shifting spheres of threat and attack. Simultaneously, scalability guarantees that the automated system will efficiently respond to an increasing rate of security events without reducing its integrity. Scope and adaptability are important in determining the automation's effectiveness in resolving evolving cybersecurity situations over a long period.

The assessment of automated compliance incident response also incorporates regulatory policies and compliance frameworks relevant to the organisation. Organisations are legally bound by data protection laws and compliance requirements which are legal obligations, as well as privacy regulations, to protect sensitive information. Adherence to industry standards and frameworks like NIST Cybersecurity Framework or CIS Controls demonstrate the maturity of automated response systems and their ability to provide efficient standardised incident response automation for compliant organisations.

To automate responses successfully, incident recovery must analyse several parameters including adaptability, precision, and multi-metric performance scrutiny. Addressing these aspects helps in gauging the reliability, in addition to effectiveness, of automated incident response systems which fortifies organisational resilience and cyber posture with proactive defences against advanced cyber threats.

Challenges and Limitations of Automation

In today's landscape, automation of incident response and recovery sits at the confluence of strategic necessity and acute efficiency need. Despite its promise, automation still comes with myriad challenges and unaddressed organisational issues when deploying automated systems for incident response, spanning from organisational structure considerations to compliance frameworks.

One of the most significant issues is automating responses to evolving threats. Cybersecurity adversaries are constantly creating and implementing new attack techniques, making the predictive ability of automated response systems a requirement. Thus, ensuring real-time automated response actions are accurate and relevant becomes a primary challenge for organisations.

Another critical weakness stems from an overdependence on specific patterns or rules set within automated systems. Although these rules may effectively respond to defined threats, they often fail to address sophisticated or novel ones, leading to erroneous results. As a result, organisations are forced to find a middle ground between detection based on rigid enforcement of predetermined rules and responsive adaptation to emerging threats.

The broad range and scale of alerts sets a high operational risk for organisations. With an increase of alerts, security teams will increasingly struggle to tell the difference between real threats and phishing attempts. This highlights the need for fine tuning automated responses to high impact incidents while disregarding non-threatening activities.

Another one of the challenges highlighted in the document is the range of issues that come with integration and interoperability. There is a lack of uniformity in practices and procedures as separate organisations are reliant on specific tools and platforms due to uniqueness in protocols and interfaces. Cross-domain automated incident response for diverse environments requires complete interoperability along with seamless integration which poses significant technical and resource hurdles.

Social obligations alongside legal domains associated with automated response actions demand scrutiny. If automated systems are unchecked, they can easily breach social norms or legal obligations. Adopting procedures to ensure governance on automation while keeping the system legal and ethical has firm consequences on the implementation of such systems.

Finally, there are risks entrenched in matters such as sabotage, undermining the trust placed in them. Technology being more advanced nowadays provides rich opportunities along with threats when it comes to automated systems. Cybersecurity is a demanding topic requiring continuous improvement as information technology has to be adaptive to new risks revolving around change. These factors lead to an automated system being sabotaged or criticised for being weak and open to manipulation.

Such matters affect governance causing greater demand for organisational readiness. Improving these matters requires an approach on the technological level while strengthening the organisation on the governance level. Response strategies revolving around automation should mitigate risks associated with cybersecurity.

Future Trends in Incident Response Automation

Considering how technology evolves, analysing the trends of tomorrow regarding incident response automation in the cybersecurity landscape is increasingly critical. The automation and orchestration of workflows to improve on existing AI-enhanced incident response frameworks is one of those trends that stands out. These platforms will implement processes and protocols based on algorithms that will learn continuously and thus sharpen their predictive analytics and response capabilities. This will enable companies to proactively defend against ever more sophisticated attacks and threat actors. The acceleration of autonomous incident response systems that are able not only to detect and contain but also to perform remediation tasks automatically are another area of advancement. While the prospect of automating the handling of security events is certainly appealing, the capricious risks that come without any form of human input looms large. AI and NLP integrations can process unstructured data like security reports, blogs, or even social media feeds, automating the response based on context, understanding, and reasoning indirectly improves incident response automation. This advancement can drastically enhance responsiveness to incidents while expediting the overall responsiveness in dealing with new threats.

The future of incident response automation revolves around the application of blockchain technology to further enhance the immutability and integrity of data involved in the incident response. Blockchain helps organisations establish secure and transparent systems for documenting and auditing actions taken during incident responses. Furthermore, quantum computing can transform incident response automation by exponentially speeding up the analysis of intricate threats, making cryptographic operations, and dealing with intricate challenges, granting unparalleled agility and precision to security teams when dealing with advanced and sophisticated attacks. As with any emerging technology, AI and automation adoption in incident response will require ethical considerations and governance attention to control their use within legal bounds. In other words, the new directions suggest a confluence of emerging technologies that will redefine, more than ever, the efficacy, efficiency and robustness of cybersecurity operations in all industries.

7
Offensive AI: The Cyber Attacker's Arsenal

Introducing Offensive AI Tactics

The implications of Artificial Intelligence (AI) technology for future offensive specialised cyber operations are unprecedented because it has already turned from a futuristic concept into reality. We are interested here in the particulars of offensive AI tactics and how one exploits AIs to bolster their attack plans and achieve malicious intents.

These days, tools and techniques powered by AI technology comprise one of the most critical powers a contender can possess. The use of machine learning, deep learning, and other features of AI enables malign actors to develop adaptive attack vectors with sophisticated AI systems that challenge even the most innovative defence strategies. To comprehend the nature of new cyber adversary AI, it is important to recognise its mind, which is key to understanding the nature of risks bolstered by cyber technology so as to improve fortification systems against these threats.

Thus far, we have explored how AI is used for cyber-attacks. Now, we focus on offensive AI capabilities by identifying specific instances of orchestrated cyber-attacks in which AI is involved. It analyses the methods Atak onua employs AI during the reconnaissance stage to derive information about target systems and pre-emptive attack entry opportunities. Its focus will also include methods employed through advanced techniques to circumvent detection through the AI evasion risk reduction techniques.

Examining various offensive AI strategies will illuminate the relationship between AI and cyber warfare. In this ever-evolving space, it is crucial to follow emerging offensive AI trends in formulating counterstrategies that would prevent hostile infringements on critical assets and infrastructures.

Understanding the Mind of an AI Adversary

While researching AI-enabled cyber attacks, one must consider the reasoning and planning processes of the AI adversary. Unlike human challengers, AI hackers operate devoid of psychological drives, be it sympathy or morality. The "mind" of an AI adversary embodies multifaceted algorithms, machine learning models, and data ingestion frameworks integrated to maximise efficiency in penetrating and compromising digital systems.

An exploration into the capabilities for autonomous decision-making and adaptive learning will lead to understanding the AI adversary better. AI adversaries have the ability to learn from their surroundings perpetually, optimising their strategies based on the responses they encounter. Because they are able to modify their tactics in response to countermeasures, and therefore, pose an everlasting challenge to cybersecurity professionals, they are dangerously efficient.

Additionally, relentless optimisation shapes the AI adversary's mindset. AI attackers are motivated by objective functions and reinforcement learning, which fuels the constant iteration of techniques to maximise success and minimise detection and mitigation. Their ability to exploit unanticipated system weaknesses, create new attack methodologies, and execute complex manoeuvres that simulate legitimate user activity is a threat to traditional defensive structures due to this optimisation-driven mindset.

Moreover, understanding AI adversaries requires understanding their particularism toward data-centric heuristics. From deep stores of relevant information, AI hackers synthesise insights which they use to mount incredibly accurate precision attacks. Through these methods, AI adversaries are able to scrutinise the cyber landscape with surgical precision, determining the best possible locations for breaches and ways to extract confidential data while evading dragnets of countermeasures with intricate calculations.

Grasping the AI adversary's mindset boils down to evaluating the adaptation, optimisation, and data-centric heuristics in AI warfare. Next, we will examine the elements that AI adversaries use to relentlessly target and attack critical digital infrastructures using advanced techniques, including the elements of AI reconnaissance.

AI-Driven Reconnaissance Techniques

The methods of reconnaissance have been transformed with the advent of artificial intelligence technology, with new possibilities of intelligence gathering and target identification emerging which greatly facilitate the attackers. By employing algorithms and analysing available data, adversaries have the capacity to perform automated scans of entire networks and information systems to collect necessary information and identify vulnerabilities devoid of human efforts. The described activities of reconnaissance performed by AI systems consist of collecting and processing data about specific computer-based environments, which makes it possible for cyber attackers to understand thoroughly the structures of their subjects of attack and ideal entry points. AI reconnaissance can be considered to surpass more traditional techniques because it allows an adversary to automate the collection and analysis of information at a rate far more rapid than manual systematic. AI enables attackers to tailor their reconnaissance to specific syndicates, employing multi-dimensional and evolving methodologies that focus on bypassing detection and counteraction. AI further enables dynamic analysis on voluminous datasets, spotting trends and drawing important conclusions which result in a disproportionate advantage to attackers as far as designing intricate attacks revolves around identified gaps.

The application of artificial intelligence in reconnaissance amplifies data gathering efficiency and assists in understanding context. This enables attackers to make well-informed decisions after thorough analyses of the gathered intelligence. Adversaries can exploit AI-driven reconnaissance techniques to uncloak attack surfaces, identify critical high-value assets, and refine offensive action schemes all while staying under the radar of security countermeasures. Any advances made in AI capabilities will be matched by new shifts in the threat landscape which makes it increasingly difficult for defenders trying to protect their systems from sophisticated reconnaissance strategies. Cybersecurity experts need to address these challenges head-on in order to proactively adapt their defences resorting to AI-based systems capable of forecasting, identifying, and neutralising risks arising from reconnaissance fuelled by artificial intelligence.

Exploiting Vulnerabilities with Machine Learning

The impact of machine learning, an aspect of artificial intelligence, can be seen across multiple industries. However, this technology can also serve as a valuable tool to cyber attackers. With the aid of machine learning algorithms, cyber criminals are able to scrutinise vast troves of data for discerning system intricacies, thereby identifying critical system weaknesses, vulnerabilities, and entry black doors bypass windows, including zero-day attacks if possible. As a result, attackers are capable of recognising patterns and out-of-the-ordinary happenings within networks and applications. Furthermore, machine learning algorithms increase in accuracy and efficiency over time as they are force-fed new data, which enables them to discover and utilise exploits with vicious precision. Moreover, the dynamic strife these algorithms constantly change with presents a hurdle to traditional defences which often depend on predetermined static systems rule-set. Unlike static systems, the dynamic adaptive nature of these algorithms can pose significant challenges to traditional cyber security defences. Moreover, the static nature of these systems makes them easy to predict. Attackers can invoke machine learning algorithms to combat the predicted outcomes, thus automating attacks exquisite self-written automated adaptations multi-faced scaled scripts and leveraged precisely timed multi-layer attacks. These approaches may mask detection on multi-layered firewalls as they tender-sensitive zero-day sneak-through VPNs in distributed businesses.

In this manner, malicious actors can design attacks tailored to bypass security measures, making detection and remediation far more challenging. Understanding the capabilities and intended consequences of machine learning when wielded by an adversary, especially in a position of power, is essential for cybersecurity experts. Informed defenders are able to formulate preemptive strategies to identify, thwart, and mitigate machine learning fuelled attacks by offensive AI advancements. This includes employing AI for protective strategies like applying machine learning algorithms to scrutinise network traffic for potential compromise indicators or flagging unusual behaviour that suggests an active attack. Also, collaboration and shared knowledge within the cybersecurity domain is imperative to stay agile against the continuously changing landscape of threats posed by cyber attacks using machine learning techniques.

Advanced Evasion Strategies Using AI

Advanced evasion strategies using AI in cyber warfare are primarily employed by malicious individuals who seek to advance industrial espionage. Modern technology offers tools that can be used for illegal purposes, but when AI is utilised, the detection gap becomes deeper than before because the systems in place to obtain intelligence can easily and stealthily be breached. AI systems can be used to design polymorphic viruses that evolve in response to environmental pressures and analyse behaviour patterns of traditional antivirus solutions and intrusion detection systems, therefore shifting their tactics to ensure maximum effectiveness. Malware is downloaded, and by employing AI systems, saboteurs have access to attack programmes that can be altered as needed so that maximum confusion is released within security systems. Moreover, AI systems allow attackers to launch polymorphic attacks which can produce many different versions of a single virus, destroying all carefully crafted algorithms and making signatures aimed at such viruses ineffective. A defining point in these strategies lies behind automated social engineering campaigns though AI machine learning algorithms. Using them, cyber criminals can analyse the habits and communication patterns of their victims to tailor very convincing phishing messages targeted at each specific individual in advance.

Such attempts at specialised spear-phishing tend to capitalise on psychological weaknesses and relationships of trust, heightening the chances of successful breaches while bypassing traditional email security systems. Additionally, the ability of attackers powered by AI to conduct automated reconnaissance and learn from previous failures enables refinement and optimisation of social engineering tactics over time. Apart from the use of AI for evasive malware and social engineering, there is increasing exploitation of generative adversarial networks (GANs) by malicious actors for the creation of advanced deepfake content intended for deception and disinformation. Applying GANs allows threat actors to manufacture hyper-realistic forged audio and video materials that enable impersonation of trusted individuals or fabrication of plausible narratives to shape public opinion or deceive targeted organisations. Such deepfakes generated through AI pose challenges for processes of authentication and validation as they can dynamically replicate authentic communications and events. While defending organisations adapt to new AI-driven evasion techniques, there is urgency for proactive frameworks that combine anomaly detection powered by AI, behavioural heuristics, and contextual intelligence to identify and preemptively neutralise evolving threats. Strengthening AI models used defensively through adversarial approaches alongside integrated threat intelligence would enable organisations to defend against sophisticated cyber evasion tactics.

Automated Phishing and Social Engineering

Cyber attackers have had continued access to powerful phishing and social engineering tools, and the application of AI technology has pronouncedly increased their impact. This will be the focus of the current section, which seeks to demonstrate how AI technology is being used for sophisticated phishing and social engineering attacks.

AI techniques are used to initiate phishing attacks by masquerading as an established communication in order to elicit a targeted response from the user. Phishing bots based on AI technology can email, text, or call their intended victims with messages that are very realistic simulations of everyday communication. These forms of communication are now very effective at bypassing normal filters used to screen emails and as a result, they pose a much greater threat than before.

AI attacks also have the tools to customise their emails more personally based on psychological triggers and emotional responses due to the extensive data collection made possible by modern technology. Addressing the target with a deep understanding of their profile using AI technologies such as sentiment analysis and behavioural profiling greatly increases the likelihood of obtaining a response, therefore resonating with the targets even more.

Simultaneously, the threat of AI integration with social engineering is profound. Automated social engineering exploits the analytical and predictive powers of AI by manipulating targets to extract crucial information or actions that would undermine security. Pattern recognition built into AI algorithms can identify target online behavioural networks as well as communication and social preferences and construct manipulative strategies to achieve desired results.

While organisations enhance their training and awareness initiatives to counter social engineering attacks, human-focused vulnerabilities are relentlessly targeted by AI-powered adversaries. AI-driven social engineering campaigns can bypass defences that rely on trusted calls and masquerade to hack through cognitive bias by employing compelling narrative arc frameworks for access.

To sum up, automated phishing coupled with social engineering based on AI intelligence poses a significant threat to cybersecurity experts; understanding these enhanced tactics is critical for devising efficient protective measures to face emerging challenges.

Leveraging AI for Credential Harvesting

The integration of AI into credential harvesting has marked a particularly malignant chapter in the ongoing saga of cyber warfare. The combination of AI with cyber insecurity has granted cyber criminals unprecedented access to meticulously organised and sophisticated campaigns aimed at unlawfully harvesting sensitive information. Credential harvesting holistically entails acquiring names, passwords, and even more sensitive personal data through automation and strategic planning fostered by AI.

One of the main methods using the power of AI to harvest credentials focuses on the exploitation of advanced phishing attacks. With the aid of machine learning, criminals can produce sophisticated phishing emails while exploiting the mental weaknesses of users. On a broader scale, phishing attempts sponsored by AI have transcended traditional boundaries and aim to change text and styles owing to the habits and likes as well as the manner of speaking to improve the chances of success in obtaining the intended credentials. Additionally, the harvesting of credentials enhanced by AI technologies also incorporates the use of social engineering. In the latter, advanced algorithms are used to analyse large amounts of data for the purpose of creating messages meant to socially engineer and psychologically manipulate humans on a psychological level. Using this technique, attackers are able to enter private groups, build social trust, and collect usernames and passwords without the knowledge of the victims, fully leveraging the basic instinct of human nature to trust people and social bonds. The use of AI technology makes these social engineering schemes easier and more effective, as these techniques greatly worsen the amount and level of effect and the rate of productivity of successful social engineering operations aimed at credential harvesting.

The deep learning models focused on behavioural analysis and advanced analytics that make predictions around user behaviour and expected need have also been integrated into the automated stealing of credentials. This enables attackers to carry out highly relevant and contextual assaults. Through the use of AI to look at huge collections of data about certain aspects of a person's digital presence, criminals are able to design specific strategies and bypass security systems effortlessly to obtain the required personal information.

The integration of AI with credential harvesting attacks presents a new set of challenges to cybersecurity AI technologies, advocating for vigilance in detection, authentication, and user awareness training programmes. With advancing technologies, it becomes imperative for security professionals to make use of AI in strengthening their cybersecurity frameworks, designing proactive adaptive counteracting measures, and building a profound insight on anticipating the ever-evolving threat landscape while defending against the consequences of AI-driven credential harvesting.

Deepfakes and Misinformation Campaigns

The advent of AI has ushered in deepfakes and misinformation campaigns, which have emerged as formidable threats to cybersecurity. Misinformation technology enables the creation of authentic-looking fraudulent videos and audio files through deep machine learning techniques, advancing deception. It poses a threat by enabling individuals to be easily deceived, manipulating social release and social order, thereby creating societal chaos. The capability to generate authentic-looking media content easily fundamentally reshapes the landscape of cybersecurity. Misinformation campaigns exploit social and technological media platforms to tell lies in order to incite violence in the community and breed suspicion among people against each other. Such organised moves have the potential to bring about disorder, destroy the currency system, and damage democracies. In cybersecurity, deepfakes can be used to impersonate important people, high-ranking officials, trusted associates, or even corporate business leaders to fraudulently obtain sensitive information or commit fraud. Moreover, the mass distribution of falsified information can destroy reputations, provoke, or kill, and cause irreparable harm to the subject's life and work. The deceptive quality of deepfake videos calls for firm and thorough controls to limit their spread. More and more organisations should focus on implementing solid technological checks and use AI to detect and eliminate tampered media.

It is essential to tech firms, scholars, and government officials that an interdisciplinary collaboration culminates in industry frameworks and policies to address the challenges posed by deepfakes and disinformation. Furthermore, public education on the existence and potential consequences of deepfakes is important to ensure social defence from their adverse impacts. Heightened alertness and readiness are crucial as the cyber threat domain shifts to the more sophisticated and clandestine detrimental effects of deepfakes and disinformation campaigns.

Case Studies of Offensive AI in Action

In the cyber defence field, case studies are very crucial in demonstrating the practical use of offensive AI in cyber strategies. These case studies focus on particular instances and campaigns that involved the use of cyber AI attacks. These instances are essential because they provide a wealth of knowledge regarding the strategies used, the outcomes, and the strategies employed for counteraction whether successful or otherwise. These case studies should not be perceived as mere narratives of past occurrences; they are instructional materials for cybersecurity practitioners and policymakers to enhance their response against attacks.

An illustrative case study is of a significant financial organisation that suffered an adverse AI machine learning attack. Using advanced techniques, the attackers bypassed traditional security measures, compromising sensitive financial information. In addition to monetary losses, the institution's reputation suffered. A thorough diagnosis provides insight into AI bypass schemes while highlighting the necessity for evolving, AI adaptive counteractive frameworks.

Another notable case study encompasses a global cyber political conflict in which a state-sponsored actor used AI to execute a disinformation programme designed to incite disorder. Using deepfake technology and tailored lies, the perpetrator caused extenuating distrust in information sources and widespread confusion. This case is a reminder of the issues that sabotage society's coherence by the use of aggressive AI technology and the urgent need to protect societies from such designs.

Moreover, analysing an AI-driven phishing attack executed on a large international company illustrates the integration of AI with social engineering tactics. The perpetrators employed AI algorithms to create highly tailored phishing emails, putting the employees of the corporation at a higher risk than usual. This situation demonstrates that no defences exist to protect organisations against social engineering attacks geared towards exploiting artificial intelligence.

These case studies illustrate the impending peril and challenging complexity posed by offensive AI within the cybersecurity domain, forcing a shift in priorities for defenders and organisations. There is a need for in-depth study of these situations so that we can learn how to better defend ourselves against attacks and put anticipatory measures in place.

Collaboration with Human Hackers

The combination of AI and human hackers marks a major shift in the cyber warfare arena. Each element has its own set of advantages and disadvantages. In this regard, it is AI technologies that continue to evolve at a fast pace and cyber warfare tools alongside them due to human skills remaining an ever greater asset in the world of hacking. The focus here is the cooperation between AI and human hackers, thus exposing the benefits, problems, and controversies of such a partnership. The collaboration of AI and human hackers gives rise to new, advanced forms of precision attacks aided by the computational ability of AI and the fluidity and instinct of human participants. The CIA Technical Services Branch calls sociological elements of hacking manipulation; human crackers possess invaluable years of experience, field wisdom, and intricate ability to deal with complicated social engineering strategies which AIs always find greatly difficult to replicate. In addition, such hackers who penetrate into the systems usually have great insight into the systems and the people working in them, hence able to devise very unique ways for unique attacks. Moreover, since human beings predict and behave in patterns, these abilities augment AI's ability to analyse large amounts of data to search for possible targets, automate some dull repetitive and menial work, and even forecast behavioural patterns of people on a very unprecedented scale. AI has its own advantages which enable it to perform these tasks much faster than a human being would. Thus, apart from these two elements, hybrid forms of the two attacks emerge, aiming to constitute the essence and contribution of both sides while reducing the losses for one party.

Nonetheless, this partnership raises issues, particularly with regard to ethics. It brings to light some of the most profound issues pertaining to AI cyber-enablement and its use in cyber-attacks, especially regarding the arms race concerning the charge and blame of autonomous systems. It also becomes crucial that AI hacking policies are defined to ensure that the actual human hackers possess total control, with respect to intelligence and responsibility, over the final decision. Furthermore, these policies must transcend AI's potential to enhance biases of human offence and maliciousness, and address the invoked issues of offensive cyber operations through AI. Given the continuous merging of human and machine-driven attacks, concern must be raised by all stakeholders and policymakers, non-profits, firms, or cyber experts and engineers to confront the laws which would appropriately allow the usage of various advances while protecting against the abuse of such advances. Addressing the ethical issues allows for the involvement of AI systems and human intelligence in attacks motivated by threats to place such mechanisms with constant vigilance, responsibility, and neutrality.

Autonomous Exploitation and Penetration Testing

Introducing Autonomous Exploitation

The past few decades have witnessed AI-driven automation revolutionising almost every domain, and cybersecurity is no different. This portion explores the primary concepts and preliminary frameworks aimed at leveraging AI for automating exploitation processes in cyber-attacks and defences. Autonomous exploitation concerns the capacity of AI to autonomously recognise and exploit breaches existing within an entity's infrastructural components. It ranges from the discovery of high-value targets to full-fledged sophisticated automated attack orchestration. The new wave of technology-enabled exploitation changes the game as far as vulnerability assessment and penetration testing is concerned with AI support. It can change the paradigm in which cyber defence strategies are structured and employed through sophisticated algorithms based on data mining, machine learning, and unstructured data analysis. This application of AI proves useful especially for counteracting exploitable weaknesses in organisational frameworks and enhancing their defensive postures even before adversaries take the initiative to level striking blows against weaknesses. Moreover, autonomous exploitation can drastically enhance the automation of strategic multi-level security evaluations within an organisation while bringing down the levels of economic spending and man-hours needed to achieve these goals.

In this part, we will examine all aspects of autonomous exploitation, including the technologies involved, ethical issues, and the implications of this paradigm shift. In understanding the autonomous exploitation paradigm, organisations stand to better prepare themselves on how to use AI efficiently as a force multiplier in protecting their digital assets within a perilous environment where threats are ever growing in complexity.

Frameworks and Tools in Automated Pen Testing

The automated penetration testing approach has changed the way companies evaluate their security posture. This part examines the different frameworks and tools used in automated pen testing and considers their advantages, disadvantages, and practical use. There are both commercial and open source frameworks available, such as Metasploit and Burp Suite Pro, which offer a wide range of services to meet various organisational requirements. Each tool has distinguishing features that provide ranging from cyber attack simulated penetration testing to exploit vulnerability scanning and post-exploitation activities. Furthermore, some new platforms use machine learning and AI technologies to automatically pinpoint and compromise vulnerabilities, igniting a new era of pen testing tools. Also, AWS Security Hub and Azure Security Centre as cloud-based services provide integrated solutions for ongoing security monitoring and compliance checks. These frameworks not only define security gaps, but also provide detailed assessments, along actionable insights to improve and remediate identified gaps. Automated tools, in as much as they are efficient, the skill level and competence of the personnel operating them determines effectiveness. Operating these tools requires a high-level skill to avoid configuration or result misinterpretations that can lead to automating false positives or negatives damaging the overall security assessment.

Besides the balance between automation and human judgment, creativity is still essential when engaging unconventional methods for penetration testing and vulnerability discovery. Therefore, an organisation's skilled professional in automation pen testing requires a delicate balance with advanced tools. There will always be room for improvement within the landscape of automated pen testing tools. As technology advances, these tools will be able to simulate complex cyber threats with deeper sophistication and agile adaptability.

Understanding the Workflow of AI-Powered Exploit Engines

The process of exploit engines powered by AI reflects one of the most critical advances in the spectrum of cybersecurity. These systems are now some of the most powerful pseudonymous systems with the ability to detect almost every form of vulnerability and automatically generate pseudonymous attacks, thus taking the burden off cybersecurity engineers.

Every AI exploit engine has in one form or another a set of systems which together accomplish some goals that make these systems effective. With regard to these engines, their workflows start with continuous observation and information recording from various levels such as system logs, application behaviour, network traffic. The input stream of information makes it possible for the engine to analyse, know patterns, and identify anomalies in the targeted environment.

As soon as relevant information is gathered, the exploit engine goes through an elaborate stage of feature extraction and data preprocessing. The system is able to extract complex information and convert it into actionable insights through normalisation and feature extraction. This will, in turn, inform the subsequent decision-making and strategising efforts.

Then, exploit engines use statistical models of possible vulnerabilities to forecast risks using preset algorithms. With the threat landscape growing in complexity, predictive models that are dynamically adjusted with iterative processes and refined continuously enhance systems' capabilities to meet evolving demands.

After identifying targets that can be exploited, the engine accesses a library of attack modules that have already been configured and uses a wide range of attack methodologies to create tailored exploit payloads for the vulnerabilities detected. By custom-tailoring attacks and dynamically changing the way they are executed, the conventional defence mechanisms are circumvented and breaches are inflicted effectively.

Throughout the workflow, the AI-powered exploit engine continuously improves its network navigation and defensive measure overcoming capabilities as it receives real-time feedback and updates. Also, the self-directing engine can learn from exploits, whether successful or failed, which enables iterative optimisation of its strategies with every attempt.

A thorough understanding of the workflow within AI-powered exploit engines enables cybersecurity practitioners to comprehend the changing environment of cyber threats and counter-cyber measures. Such understanding arms defenders with the knowledge and capabilities to design sophisticated proactive defence mechanisms that can withstand the power of AI-enabled cyber attacks.

Algorithmic Innovations in Automated Vulnerability Discovery

In the modern landscape, proactively identifying security weaknesses is critical to enable organisations to block potential attacks to mitigate damages. In the past few years, the development of automated systems designed to detect vulnerabilities in a wide array of digital assets has greatly increased due to new algorithms. Our focus is on the different algorithmic strategies and methods concerning automated vulnerability discovery.

One of the foremost machine learning approaches to automated vulnerability discovery is focused on the implementation of machine learning algorithms for anomaly detection. Such algorithms, by virtue of their ability to process large volumes of information, can flag anomalies that would otherwise escape notice by standard scanning tools as potential vulnerabilities. Moreover, with the introduction of evolutionary principles in some algorithms, the concept of automated vulnerability discovery, as well as advanced scanning techniques, has greatly benefited from these algorithms which mimic the process of natural selection in refining scans over time.

With the aid of algorithms based on heuristics, automated exploration of complex programme architectures to identify potential weaknesses has been made possible.

These algorithms provide automated systems the capabilities to collect pertinent information from vast codes and infrastructures, as well as detect exploitable weaknesses critical to the cyber posture of an organisation and the risks they pose. Additionally, advanced algorithms for pattern recognition also contribute significantly towards detecting known threat ranges by recognising repetition of elements and efficiently scanning and classifying potential threats.

Along with improvements in algorithms, applying threat intelligence and contextual reasoning to automated vulnerability discovery processes has improved the efficacy of such systems. Automated systems can now utilise real-time threat feeds and historical data of attacks to assess the likelihood of a vulnerability being exploited, thus allowing an organisation to focus on areas requiring the most attention.

With the development of new attack vectors, maintaining focus on algorithmic innovations in automated vulnerability discovery is crucial. The landscape of technology is shifting owing to persistent innovations aimed at enhancing automated systems' capabilities to identify and address vulnerabilities with the use of advanced algorithms, including deep learning approaches for code semantic understanding and proactive fuzzing algorithms that mimic real-world cyber threats.

Risk Assessment and Impact Analysis Using AI

The process of evaluating and analysing risk in the context of cybersecurity is essential in protecting digital assets from any malicious threats. Using AI for impact analysis and risk assessment represents a fundamental change in the organisational defence strategies. AI systems can utilise machine learning algorithms, as well as large amounts of data, to scan for system logs, network data, and other pertinent information to identify possible gaps within an organisation and evaluate their impact on operations. AI systems can also monitor and analyse evolving threats to an organisation and adapt in real time, thereby ensuring accurate assessments and enabling security teams to address the most critical risks.

Through historical data combined with current threat landscapes, AI systems can identify new attack vectors and thus allow organisations to bolster their defences. Lastly, AI technologies can allow the real-time adaptation of strategies for mitigation of risks arising from changing scenarios which improve postures in organisational cybersecurity agility.

Furthermore, AI-enabled risk assessment tools can offer an organisation unparalleled visibility into how assets are interconnected and their value, as well as how a breach of security may impact them, enabling proactive preparation and resource allocation for response and recovery efforts.

AI-enabled impact assessment and simulation features allow security teams to perform "what-if" queries for various attack scenarios and assess possible outcomes to enhance incident response planning and preparedness. Organisations can estimate possible costs and cascading effects of security incidents to better decide on resource allocation and the prioritisation of response actions. Addressing ethical issues and biases embedded in algorithms utilised for risk assessment and impact analysis is critical, though. Applying appropriate control mechanisms to ensure transparency, fairness, and accountability in the use of AI technologies for risk assessment is essential for preserving trust and ethical credibility in cybersecurity. As AI evolves, its application will greatly enhance the organisation's ability to foresee risks, scan the environment for imminent danger, act without delay, and swiftly address emerging cyber threats.

Integrating Machine Learning in Exploit Strategy Formulation

Gaining access to and properly exploiting the vulnerabilities within a secure system requires unrivalled tactics that guarantee outsmarting advanced defence approaches. Exploit strategy formulation has machine learning integration as one of the most important recent developments in the cybersecurity field. We are going to discuss how ML is influencing the pushing of dynamic and effective exploit strategies in depth.

The concept of exploit strategies developed based on learned methodology evolves in response to adjusted capital. Combining ML with strategy formulation enables swift analysis of massive quantities of data to uncover interwoven complexities that are mute to traditional methods. System security professionals will be able to identify evolving trends and emerging potential weaknesses utilising ML driven algorithms, thus enabling formulation of more precise exploit strategies.

Adapting exploit strategies to exploit formulation using machine learning models requires constant monitoring of evolving vectors aimed at executing diverse attacks. Through the continuous monitoring, ML models are able to safely navigate around newly integrated security protocols, behaviours, and defensive measures. Adaptability to lessen offensive techniques and easily bypass progressive fortified security architectures improves through this strategy.

The use of ML intricately allows for proactive detection of zero-day threats and likely pathways for exploitation. Past incidents and abnormal patterns within the data can be analysed and exploited within systems, allowing for mitigation efforts to be activated in advance whilst defensive measures can be reinforced ahead of time.

Additionally, the context-aware exploit strategies can further be developed through the use of ML, raising their overall precision. It can be tailored to a particular environment and with the real-time situational feedback adjustments, a minimal risk of false positives can be ensured.

On the other hand, formulating exploit strategies integrated with ML presents challenges such as lack of quality training data, potential attacks from adversaries on ML models, and autonomous actions taken ethically without reasoning. These hurdles need to be resolved in order to fully take advantage of and shape future exploit strategies using machine learning.

To conclude, security specialists developing adaptive strategies find that ML offers them a chance to pivot with fine-tuned precision as proficient cyber tools to navigate through the multilayered security landscapes.

Challenges and Limitations in Autonomous Exploitation

Although autonomous exploitation can enhance penetration testing, it poses challenges and constraints that need attention. The vast complexities of the cybersecurity ecosystem are one of the major issues, especially with multifaceted problems like threats and breaches.

Autonomous systems do not yet possess the necessary contextual understanding and adaptability. Furthermore, the multitude of network domain topologies and interdependencies among varied technologies tend to be quite challenging for autonomous exploitation systems to overcome.

Fragmented and outdated intelligence on emerging threats can affect the accurate determination of exploitation prerequisites, nothing its precision. Automated exploitation faces another major challenge with positive and negative errors. Detection and prevention systems operating with established static standards will miss real dangers, resulting in reduced trust confidence and accuracy of autonomous systems. To a lesser extent, ethical issues are incredibly important with autonomous exploitation. Automated systems producing autonomous attacks can lead to unintended network collateral damage, raising crucial ethical issues that require significant exploration and mitigation.

Maintaining trust in the cybersecurity community hinges on ensuring adherence to legal and ethical exploitation boundaries when using autonomous systems.

The autonomous systems' self-exploitation domain is still under regulatory development, which brings both difficulties and ambiguities for the system's creators and users. With the growth of autonomous self-exploitation, compliance with emerging policies becomes increasingly convoluted. Furthermore, a human dimension must be considered when discussing self-exploitation concerns. While machine learning and AI technologies may improve automation in many penetration testing tasks, the persistent problems undermine fully automated self-exploitation. Such issues include the ability to grasp context, comprehend the wider implications of some actions and decisions tailored to the situation, which all remain monumental hurdles for automatic capability. These concerns and constraints impact the autonomous self-exploitation cellular system.

Ethical Considerations and Legal Implications

With the shift to autonomous exploitation and penetration testing, ethical considerations and legal implications, especially in the context of AI-powered tools, are increasingly important. Automation using artificial intelligence (AI) technologies is already making headway in the various sectors of cybersecurity. It raises ethical questions when autonomous systems are used for offensive operations. The potential of AI to identify, exploit, and compromise vulnerabilities so swiftly creates complicated moral problems and ethical dilemmas that call for scrutiny. Ethical considerations cover all issues concerning the impact of autonomous cyber exploitation and penetration testing on individuals, companies, infrastructure, as well as their effects on world security and stability. Such deliberate ethical exploitation strategies require policies that take into account ethical principles such as responsibility, transparency, accountability, and collateral damage minimisation in aggressive cyber maneuvers. Ethical issues related to autonomous exploitation are part of a broader cyber regulatory issue which is complex and intersects with legal issues.

Liability and jurisdiction issues, as well as the operational boundaries of international laws and norms, stand out as primary considerations in dealing with AI-driven offensive operations. The continuing conversations at the intersection of law, policy, and technology are crucial in addressing the sophisticated legal issues posed by autonomous cyber systems. The cybersecurity community can explore existing legal approaches and advocate for delineated policies to reduce legal uncertainty and guarantee that the use of autonomous exploitation and penetration testing is legally compliant. Strong collaboration between all relevant actors, such as policymakers, jurists, and computer scientists, is needed to create an appropriate framework where technological development can both respect and be integrated with legal frameworks and social ethics. This proactive ethical and legal stance affirms the use of AI in cyberspace defence operations while maintaining integrity, trust, and fundamental human rights.

Comparative Analysis: Manual Versus Autonomous Methods

A traditional system security assessment approach

called manual penetration testing relies on human testers to follow processes designed to identify weaknesses within a system. Testers do most of the resource-heavy processes because this approach is very tedious and relies significantly on their experience. On the other hand, autonomous methods seek to improve penetration testing processes by incorporating elements of AI and machine learning. In this part, both methods are discussed in depth including their analog and digital similarities and differences.

While effective, manual penetration testing is hampered by the inability to adapt to evolving challenges dynamically, scalability concerns, or even human inaccuracies. Also, information gathering may not be exhaustive as a result of time constraints or lack of oversight. Autonomous methods address many concerns related to manual testing because they allow for continuous testing, swift detection of vulnerabilities, scaling to complex large systems, or even constructing elaborate frameworks to resolve issues.

The manual approach to vulnerability assessment and analysis is arguably one of the most tedious methods of operation as it requires human input. Since automation seeks to eliminate the need for human input by completing tasks faster—scanning massive networks for gaps within the set parameters—autonomous strategies dramatically outperform manual methods in both speed and efficiency.

Additionally, manual testing methodologies do not do well with keeping pace with an ever-changing attack surface. In contrast, autonomous approaches are able to learn from new attacks even as they are being integrated. The proactive changes allow for greater adaptability on the entire system's architecture and resilience on the entire system.

When making comparisons between techniques, impact, and overall benefit to an organisation, economical efficiency stands out. Thorough assessments performed by competent experts tend to drive costs through the roof. Advanced operators will enable organisations to redirect their budgets towards other needs because, after the initial setup, such operators work on automated cycles with little human supervision.

Both strategies as a whole contain their own advantages and disadvantages. Manual assessments allow for creativity, intuition, and contextual understanding that is difficult to programme into automated frameworks; in contrast, autonomous techniques focus on agility, rapid execution, and seamless scalability.

Ultimately, the discussion on each method's advantages and disadvantages has revealed some remarkable differences between manual and automated penetration testing techniques. The cybersecurity practitioner is better placed to determine which method best augments the defence of a digital system with timely intelligence on emerging threats if they appreciate the strengths, weaknesses, and differences of these methods.

Future Directions and Emerging Technologies in Autonomous Penetration Testing

The anticipated future of autonomous penetration testing is promising, particularly at the junction where artificial intelligence intersects with cybersecurity. Because of the persistent evolution of technology, new threats and vulnerabilities will be continuously emerging. Thus, ML algorithms are being created that will work towards greater adaptability of autonomous penetration testing systems to such changes, which will ultimately make it more efficient in targeting exploitable weaknesses in the system. Moreover, the application of sophisticated contextual and threat intelligence to autonomous penetration testing tools promises to transform the industry. These techniques not only analyse information but also use predictive models to uncover gaps and anticipate complications before they occur. This will help organisations avert cyber threats as well as reduce the risk of security breaches by proactively dealing with issues.

Another emerging trend is the application of blockchain in autonomous penetration testing. The structures and attributes of blockchain, such as decentralisation and its immutability characteristic, facilitate robust safeguarding of records in the penetration testing procedure. This feature enhances accountability and reliability by ensuring that all test results are protected against unauthorised modifications and thus, confirmed to be authentic.

Quantum computing has recently become an option and challenge at the same time for autonomous penetration testing. They can solve numerous complex problems which, for a long time, could only be addressed by traditional supercomputers at unimaginable speeds and at once, especially cryptographic problems. Therefore, autonomous penetration testing systems need to adopt and strengthen their cryptographic defences against quantum-powered rivals.

I find the intersection between autonomous vehicles and penetration testing intriguing as well. With the rise of IoT devices and systems, there is a need for a security assessment in more unconventional areas like the smart city and the connected infrastructure ecosystems. Autonomous penetration testing systems designed for these environments will be crucial as we strive to design security for the infrastructures and ecosystems of advanced technologies.

To conclude, the development of autonomous penetration testing is heavily anchored to the constant technological advancements. The adoption and integration of new technologies— including artificial intelligence, sophisticated threat intelligence, blockchain, quantum computing, and the Internet of Things— will further strengthen the autonomous penetration testing framework, enabling it to effectively defend the digital ecosystems against sophisticated attacks.

9
Case Studies: Battles in the Wild

Real-World AI-Driven Cyber Incidents

The rapid spread of artificial intelligence tools and technologies has offered unique opportunities for businesses and individuals. At the same time, it poses new threats with its ability to transform the nature and type of risks in the cybersecurity situation. Today's global trends in digitalisation and the growing volume of international traffic have intensified cyber risks. In this part, we will consider the general questions of implementing AI in cybersecurity in more depth and analyse AI-initiated attacks from different angles.

In the face of escalating cyber complexities, these developments only foreshadow a new global race for AI-driven security applications. Global organisations and authorities have been slow to take real initiative and protect their infrastructures. It also makes the analysis of incidents much more important, since the methodology covers numerous different aspects which are usually overlooked when worrying about more theory-based approaches. These attacks reveal artificial intelligence-driven cyber risks that require protective technologies due to their disclosure.

Beyond threat detection, the application of AI in cybersecurity also includes automated response sequencing, anomaly detection, and forecasting. Such capabilities provide an organisation with agility and strategic foresight to counter ever-evolving threats. Knowing how

AI interacts with cyber events is crucial for developing robust multi-layered defence architecture that can adapt to continuous changes in adversarial tactics. The strategic application of AI as a force multiplier enables optimised defensive operations and thus warrants an in-depth study of AI-activated cyber incidents. This analysis not only aims to clarify the complex connections between AI and cybersecurity but also seeks to highlight the need to utilise AI proactively in cyber defence. Framing this discourse within the larger context of cyber warfare aims to inform stakeholders and empower them with relevant intelligence to strengthen their posture against AI-based attacks. Assimilating AI will help neutralise risks to critical systems while established digital front lines will aid in safeguarding assets during an age of rapid technological advancement and unrelenting cyber hostilities.

Case Study: The Financial Sector Siege

In the context of contemporary cyber warfare, the siege on the financial sector is particularly noteworthy with respect to the sheer scale of AI-enabled attacks. The sophistication of such attacks obliterated established systems of defence, creating a new reality for financial institutions and banks globally. Be it first DDoS attacks or systematic data breaches, the onslaught was coupled with intelligent algorithms designed to capitalise on weaknesses, AI technology was deployed in every strategy to dismantle the financial networks. This is a multi-faceted case study that explores the intricacies of this siege while detailing the profound impacts of the attack on the global economy and regulatory frameworks, markets and legislations. With surgical precision, we analyse the timeline of events which reveal the systematic economic disruption and destabilisation that was the motivation behind the attacks. Even as we trace the elaborate tapestry of infiltrations and incursions, we strip off layers of anonymity veiling the attackers who were remarkably agile and resourceful. They adapt unlike anything that has been seen before. And finally, we draw attention to the urgent need for preemptive defensive action and resilient strategies based on AI technologies in the face of protective postures highlights the alarming nature of the threats posed.

This case study offers in-depth analyses and describes the insights and strategies critical to the financers, policymakers, and cybersecurity experts, culminating in a proactive approach to enhancing cyber defence mechanisms and deterrence strategies. While stressing the importance of joint collaboration to withstand the assault on the financial sector, we highlight the need to unite diverse information and advanced technologies to strengthen impenetrable defences that protect against relentless attacks.

Breaking Down the Healthcare Heist

Healthcare organisations are now a rich target for cyber attacks focusing on advanced persistent threats (APTs) in healthcare networks as a result of the critical nature of data. The integration of healthcare and information technologies has created a new battleground for healthcare defenders and attackers. This is where we analyse one of the notable cases of AI powered cyber cancer which infiltrated the AI Power healthcare networks. The case in point involved sophisticated AI algorithms that infiltrated the organisation's systems and penetrated the patient records while compromising essential medical services. AI-driven viruses were used to encapsulate bypassing firewalls and traditional security measures embedding within the network structure. This allowed them to deeply implant themselves thus systemically gaining access to vital information while masking the data encryption standard. The techniques the attackers used were quite illuminating opening and converge not only AI and Healthcare, Cybersecurity, but Cyber Warfare as well. Moreover, we examine the impact pervasive breaches of privacy expose patients and compromise the integrity of medicine while simultaneously undermining social trust within the public healthcare system. This dissected document opens new ideas of strategically defining the diverse intercontinental approaches of cyber healthcare warfare AI Systems.

Analysing AI-Powered Attacks on Critical Infrastructure

The attack vectors involving AI technology become clearer when examining the critical infrastructure components that are interdependently proprietary. This part looks more thoroughly into the complexities of such attacks and their prospective impacts while suggesting some possible countermeasures. Permeable cyber-physical systems in charge of vital services like energy, transport, and telecommunication are the most accessible levels for sophisticated AI-enabled attacks. The potential impact of AI integration into malware including but not limited to adaptive and self-modifying AI programmes pose serious risk to the functional health of these infrastructures. Recently, and through deep investigation, we have begun to confront bold realities owing to the fact that AI's agility combined with its learning capabilities enables it to commit and learn new disruptive tactics on infrastructures.

Looking further into these matters, one faces the multi-layered difficulties of countering and identifying such attacks. AI social engineering makes phishing attacks much easier to perpetrate due to the ease of manipulating humans' sophisticated system. These AI-powered breaches suffered from one unique feature that makes the problem of masquerade attack much more pronounced, namely stealthiness. Solving these problems requires developing out-of-the-box methods that combine cutting-edge anomaly detection, behavioural analysis, and GPU-accelerated threat intelligence.

The proposed cascading impacts of a successful breach on critical infrastructure systems need in-depth analysis. Beyond services being temporarily disabled, the secondary impacts that include economic losses, environmental damage, and threats to public safety raise the need for effective risk management strategy and crisis response protocols. Based on case studies, this part illustrates the complex dependencies and relations between different sectors as well as the need for collaboration in building defences against AI-driven attacks.

Fundamentally, the call to action presented in the analysis can be viewed as a shift towards infrastructure protection. It calls for the design and development of customisable defence systems made accessible via advanced artificial intelligence infrastructure. A shift from traditional models of anticipating threats and attacks towards investment in AI-driven motivation analysis, incident forecasting, strategic scenario simulation trainings, and cross-sector collaboration is necessary. Structured approaches on the issue such as establishing laws or rules and inter-institutional frameworks to deal with the AI-enhanced threat, will enable protection against the consequences of an assault targeting the very foundation of society.

Lessons from the Corporate Data Breach

A corporate data breach is not only a catastrophe for the organisation being targeted, but also serves as a sobering reminder of how the complexities of AI and cyber defence systems intertwine, and where their gaps lie. Such a monumental breach revealed critical gaps in the security systems of contemporary businesses, creating a wake of urgency aimed at industry leaders as well as cybersecurity professionals simultaneously. As we start to peel the layers off of this unsettling incident, we see the myriad lessons that resonate throughout the digitally connected world.

Primarily, this illustrates the ever-elusive and growing danger of coordinated efforts seeking to steal invaluable business secrets, especially exploiting sophisticated AI infrastructure systems. Affected organisations have paid dearly as the defences deployed and relied upon crumbled like a house of cards. Its impact though, reached further than data, tearing apart brand loyalty and trust in a business.

At the same time, the breach forms a compelling case study about the adoption gap around the continuous evolution and vigilance cyber practices need. It also highlights the importance of employing advanced threat intelligence systems capable of harnessing AI for proactive threat detection and neutralisation. Besides, this breach served as a reminder of the need to reinforce the security architecture with advanced anomaly and behaviour monitoring systems to better defend against evolving attack vectors.

In summary, the data breach inflicts a stark reminder of how deeply AI, cybersecurity, and organisational resilience intertwine. It drives the rethinking of the defensive angles used, thus calling for the use of a habitually flexible security approach that adapts to changes and employs AI-based solutions to outsmart adversaries. In essence, the main lessons drawn from this event reinforce the need for more coordinated, multilayered strategies to protect critical assets from aggressive cyber attacks.

The Rise of AI in Nation-State Cyber Ops

AI's incorporation into cybersecurity has become increasingly important for nation-state cyber operations. With AI serving both offensive and defensive purposes, these operations have become more refined and detailed than they once were. AI is being utilised by nation-states to bolster capabilities in surveillance, espionage, and sabotage, thus practically redefining cyber warfare.

AI is changing the game for nation-state cyber operations by automatically retrieving and processing huge datasets. The speed and precision at which AI systems scan data enable cyber operatives to pinpoint weaknesses, tailor specific attacks, and conduct complex reconnaissance missions with unparalleled, previously impossible levels of efficiency.

Cyber operatives can harness AI to simultaneously change and optimise many of their cyber offensive and defensive strategies as situation-specific threats evolve. Using algorithms based on machine learning, nation-states may create malware that adapts itself to evade detection and bypass security measures, thus significantly complicating matters for defenders. AI also enables attackers to execute more complex, coordinated attacks on critical infrastructure, financial systems, and government networks, thereby increasing the potential scale of damage.

As a result, the integration of AI with other, newer and developing technologies, like blockchain and quantum computing, further enhances the effectiveness of nation-state cyber operations. Attacks powered by AI can take advantage of the vulnerabilities that are typical of these modern technologies, which in turn creates an asymmetric advantage for attackers, thereby making the work of defenders more difficult when it comes to protecting critical assets and national security interests.

The soaring incorporation of AI into cyber operations of nation-states requires a complete overhaul of conventional defence frameworks. Governments, alongside cybersecurity professionals, face a mounting challenge to create effective mitigative and defensive strategies capable of dealing with the emerging threats fuelled by hostile state actors using AI. Therefore, it calls for the urgent need for a blended approach stemming from the collaboration between the public and private sectors to strengthen the cyber perimeter, foster shared threat intelligence, and build a united stance against adversarial AI exploitation in cyberspace.

With forecast perspectives, the use of AI in cyber operations of nation-states accelerates the need to focus on international legal norms, treaties, and regulations around the ethical and responsible employment of AI technology within the realms of cyberspace. Such comprehensive international frameworks would help avoid the risks arising from the arms race of these technologies and, in turn, offer relative peace and security in the cyberspace domain.

Patterns and Tactics of AI-Based Ransomware

We are now living in a world where cyber threats magnificently continue to emerge on a mega scale. Security experts are facing the most challenges when it comes to ransomware attacks due to the rapid growth of artificial intelligence. Such ransomware is AI powered; it operates with unprecedented agility and adaptability, which makes it impossible for any old-fashioned defences to succeed. All the patterns and tactics of AI-based ransomware will be discussed in detail to inform how exactly it works and the many ways it is impactful. Combating ransomware with AI algorithms enables more sophisticated AI-enabled ransomware that can self-adjust to the disposed heuristics of the environment. These advanced algorithms adapt in real time to the situation, aiming at critical information and can quickly encrypt documents, giving almost no time to access them, therefore demanding. It is even more beneficial for perpetrators turned intelligent agents able to spy, masquerading as low-profile agents and exploiting weaknesses for high multi-layered target approaches. Because of the attack focus, further damage to wealth can be obtained. Using AI-enforced ransomware architectures, the crime does not have to gain access after simulating lock-step traffic scans while the users can access networks through undetectable back doors. This imposes yet another layer to the problem. It grows greater; AI-charged ransomware is not bordered with user simulation. AI challenges aplenty, containment ex blames utilising deep-reaching resources the difference between borderline genuine and bane deeds denying torrents. Furthermore, the intersection of AI capabilities with ransomware not only strengthens offensive tactics but also enables the creation of tailored variants for particular industries, with custom threats having distinct attack vectors designed to exploit sector-specific vulnerabilities. From every stratum, organisations need to understand that the patterns and tactics of AI-based ransomware evolve continuously, and they need to take proactive steps by using AI solutions to counter these dynamic threats. Also, collaboration within the community should be encouraged to counter these threats by sharing intelligence and insights, as it helps combat the agility and potency of AI-driven ransomware. This exchange can address the refinement of the active response, mitigation response, and anticipatory responses to lessen the impact of AI-based ransomware. As generations change, the battlefield in which people live evolves; so do the techniques used. Therefore, having understanding and the ability to adapt becomes essential to combat more effectively using AI-driven ransomware.

AI in Action: Defense Mechanisms in Key Battles

With the rise of AI in cyber warfare, advanced defensive strategies have emerged to deal with cyber security threats. In the conflict involving AI-enabled ransomware, various attack mitigation and avoidance strategies have been implemented by different sectors. One critical method involves the use of AI and machine learning anomaly detection and behaviour monitoring algorithms. These systems can streamline user and network behaviour data and identify abnormal activities such as ransomware operations, suspending them.

Automated access and control restrictions, along with proactive network segmentation, improve security by inhibiting the spread of ransomware throughout other network segments. Active defence approaches employing collaborative AI systems to share behavioural threat indicators improve collective cyber resilience. Newly developed methods such as homomorphic encryption and secure enclaves present innovative ways to shield sensitive information from ransomware encryption operations. Using AI and historical data on cyber-attacks enables predictive threat modelling for ransomware deployment, allowing targeted deployment of defence measures. Besides, ransomware infection containment and recovery automations can be executed using response playbooks powered by AI.

These defence mechanisms require automation prowess, along with integration with an all-encompassing risk management and governance structure. With AI technologies influencing decision-making, security team communication and cooperation become integral in implementing a unified defence plan. As artificial intelligence enhances the capabilities of defenders, the frontline defending against ransomware and other cyber attacks is simultaneously more heavily guarded. The integration of human intelligence and AI enhances the flexibility and strength of defence systems, creating a strong shield against increasingly sophisticated cyber attackers.

Outcome Analysis and Aftermath Management

Following AI-enabled cyber incidents, it is necessary to perform outcome analyses to determine the scale of damage done, how effective the cyber defence strategies were, and whether or not AI systems employed during the incident proved useful. Such analysis helps organisations learn from the incident and update their defences for future confrontations. We cannot ignore the fact that effective "aftermath management" is just as critical as the active response in an organisation's overall cybersecurity posture. Effective aftermath management begins with a systematic post-incident briefing examining the attack's timeline, the role played by AI frameworks during the incident, and the choice actions of the human operators. By analysing these details, organisations stand to pinpoint additional weaknesses in their defensive frameworks along with appropriate improvements. Engage in comprehensive impact assessment alongside the gaps analysis to extend this effort further. An AI system or any other tool needs to work towards mitigating every form of cyber damage. Knowing how the cyber incident impacts the organisation helps quantify damages and loss incurred, estimate recovery expenses, determine reputation loss, and loss of cyber-insurance resulting in automated recovery prioritisation. Along with the previously mentioned details, ongoing AI-enabled systems need to continue evolving based on the lessons from the incident.

The information obtained from the attack should be used to improve the AI algorithms so they are able to detect and adapt to new threats more efficiently. Additionally, the interaction of human analysts with AI should be studied in order to improve the collaborative techniques that involve both domains. While drafting the response strategy to the aftermath, crisis communications take centre stage. Organisations must communicate proactively and transparently with stakeholders, which includes the customers, partners, and regulators to control the damage caused by the incident. During these sensitive periods, communication needs to be centred around clarity, expressed honesty, and shielding actions taken to correct the issues. Legal and regulatory obligations remain unchanged and should be followed after the incident. There needs to be adherence to laws concerning the notification of data breaches, the reporting of incidents and any subsequent audits or investigations that may arise. Furthermore, policies and regulations pertaining to the relevant matters must be formulated and refined based on the insights garnered from the incidents. While navigating through the shifting scenarios of AI-empowered dilemmas of cyber conflict, the relative ease with which organisations manage cyber incidents will indicate their resilience and adaptability. Differentiating the leaders from the followers, pioneering this field will be the centre of future discussions on AI-empowered cybersecurity.

Insights and Projections from Case Studies

It is valuable to extract insights and anticipate relevant trends in the field of cybersecurity after AI-related cyber incidents and their consequences, as discussed in the previous chapters. These case studies not only provide clarity on sophisticated AI-driven attacks, but also grant tremendous foresight regarding the future face of cyber warfare.

The cumulative analysis of these case studies depicts the rising sophistication and effectiveness of AI-related cyber-attacks. It is apparent that cyber adversaries are applying deep learning and automated technologies to execute multi-dimensional and stealthy systematic infiltrations capable of heavily compromising essential systems and infrastructures. There is now an urgent need to adapt defensive strategies and countermeasures to effectively deal with the new advanced persistent threats.

Moreover, these case studies illuminate a more comprehensive perspective regarding the impact and impending consequences of cyber attackers focusing on financial institutions or launching devastating ransomware attacks on healthcare networks. Businesses across the globe are continuously succumbing to harmful cyber attacks as the trend focuses on widening the scope of impact. This becomes all the more critical considering the ever-changing landscape of emerging threats. This clearly highlights the need for adaptable and proactive adjustments to security infrastructure.

Integrating AI into cyber warfare, as we extrapolate from the provided case studies, prompts the most profound ethical and legal dilemmas. The legislated offensive capabilities that are done automatically raise questions regarding who wields authority over them and consequently highlight the need for immediate legislation regulations and international coalitions to tackle the rapid growth of issues.

Furthermore, these case studies aid in recognising the threatening need for collaboration and expansion of cross-sector cooperation as well. When combined with advanced persistent threats, AI technology presents a whole new challenge for devising effective AI-powered defence structures. In today's highly interconnected digital environment, action from all stakeholders including industry players, governments, and cybersecurity professionals is vital.

The conclusions drawn from the comprehensive examination of these case studies highlight the importance of AI concerning modern cyber threats. From these lessons learned, forward-thinking approaches can be developed to mitigate the forthcoming AI-assisted cyber threats and strengthen the digital society as a whole.

10
Ethics, Law, and Policy in AI-Powered Cyber Warfare

Defining Ethical Boundaries in Cyberspace

The arena of cyberspace is filled with pivotal problems and questions of an ethical nature that have a substantial impact on international security. With the increased adoption of artificial intelligence in cybersecurity, there is so much more urgency to establishing and maintaining ethical frontiers. At the heart of the inquiry rests the exploration of ethical norms, which in one way or another structuralise action in cyber warfare and have a bearing on state and non-state actors. The connection between technology, ethics, and international security reveals the obligation to examine the ethics involved in cyberspace operations, in light of the need to study the principles of proportionality, distinction, and necessity, paying attention to their application in a globalised and digitally interdependent environment. Besides, the issue of sovereignty in cyberspace raises difficult ethical issues concerning the management and protection of state boundaries as digital territories, offensive artificial intelligence and its effects on geopolitical stability. Autonomous artificial intelligence systems present ethical problems regarding their capabilities and limitations, especially in making decisions within infinitesimal timeframes that could have drastic effects. This raises issues of ethical supervision and responsibility frameworks required to manage these technologies in a socially acceptable manner.

As has been noted, setting moral parameters for cyberspace necessitates contacts with policymakers, lawyers, technologists, and ethicists at all levels as they try to balance AI, cybersecurity, and international relations on one hand, and global ethics on the other. In the end, we will focus on the moral issues of cyberspace by analysing the ethics of AI-enabled cyber warfare and its relation to global security.

The Legislative Framework for AI in Cybersecurity

Cybersecurity and the AI intersection require clearly defined legislation as they both utilise and deploy issues of data management and privacy rights responsibly, keeping in mind responsibility and disclosure. Cybersecurity AI systems must be policed by laws regarding the protection of data and privacy compliance algorithms so legislators must think through the aspects of information privacy laws. The legalities regarding collection, retention, and utilisation of data through algorithms must be outlined keeping the sensitivity of threat affiliation in mind, and laws pertaining to information must be followed. Amidst improving the system with precision and accuracy, AI collaboration with humans creates a paradox of where the real problem lies and diverts attention away from responsibility toward artificial intelligence devoid issues. Evading control technology place needs careful monitoring as the legislators decide the depth of autonomy motioned toward AI systems. To ensure protection against cyberattacks, AI algorithms and models developed for monitoring cybersecurity necessitate precise bounds on predictability through the legislation guideline.

The ability to comprehend the rationale behind AI-driven decisions is pivotal in building trust and accountability both in the organisational context and society at large. It is the role of the policymakers to define the boundaries of explainability and auditability of AI systems to enhance their transparent governance and ethics. At the same time, policy is needed which considers the transnational aspects of AI concerning cybersecurity. Due to the nature of cyber threats, there is a need for cooperation between states and a unified approach. This requires the resolution of diverse legal issues and uniform policy in many jurisdictions regarding collaborative fighting against cyber warfare. Such treaties and agreements could set fundamental frameworks and encourage state cooperation on the common problems of artificial intelligence in military defence. In summary, the policies on AI in cybersecurity should be proactive and reactive—to shift faster than technology moves, while steadfastly upholding ethics, principles, and integrity of the digital environment. It is possible for policymakers to bridge technological prospects of AI and risks associated with it by nurturing responsible governance, thus, strengthening cyberspace security and resiliency.

International Law: Navigating Global Cyber Norms

From the perspective of international law and global cyber norms, the most pressing issues are becoming increasingly evident within the context of cybersecurity. Cyber issues transcend physical boundaries, necessitating a universally accepted legal framework for dealing with transnational cyber threats. The intricacies involved in the attribution of cyber attacks to particular actors or entities in different jurisdictions illustrate the need for international collaboration and agreements.

Presently, there is a lack of international consensus on cyberspace law, as countries continue to face the dilemmas posed by cyberwarfare. Attempts to establish cyber legal frameworks are complicated by varying perceptions of sovereignty, jurisdiction, and appropriate responses to cyber incidents. Furthermore, the absence of consensus in counter-cyberspace law results in disparate responsiveness to aggression and counter-aggression, creating inconsistency in enforcement action and diplomatic backlash. In spite of these challenges, there is significant progress being made towards international cyber law cooperation, such as the Tallinn Manual and the Budapest Convention on Cybercrime.

The introduction of cyber capabilities powered by AI technology, however, raises new ethical and legally defined questions affecting the international stage. The use of AI technology in cyber warfare creates problems in responsibility, openness, and automated warfare processes targeting systems of offence. Furthermore, the concerns regarding compliance with international law and rules of engagement in military conflict where AI-enabled cyber warfare is involved remain ambiguous. Therefore, the integration of AI technology into global cyber frameworks requires extensive discussions and agreements between states. This means dealing with issues like the autonomous offensive engagement of AI systems and defining ethical boundaries concerning the military application of AI technologies. Mechanisms for accountability regarding AI active cyber incidents also have to be formulated. The issue of international cyber law framed AI warfare will entail the enduring constructive work of the global community to address the escalating consequences of autonomous systems.

The Role of Government and Policy-Makers

As the landscape of AI cyber warfare continues to evolve, government and policy-makers take on an increasingly vital role in addressing the ethical and legal challenges posed by AI technologies. The government is responsible for formulating policies that not only guide the development and implementation of AI in cybersecurity but also serve the needs of the nation.

Balancing national security needs with protecting individual rights, privacy, and international standards is a complex responsibility for governments and policy-makers. To maintain this balance, collaboration between the government and industry, academia, and even global partners is crucial in developing frameworks that mitigate risks to AI technology while stimulating innovation and growth.

The first focus when dealing with advanced technologies and the cyberspace domain should start with governments. They need to identify the boundaries of cyberspace offensive actions, define the scope of data collection and its utilisation, and delineate rules for dealing with cross-border incidents. They must also draft a policy for the establishment of ethics committees or other appraisal divisions which are charged with monitoring the unlawful use of advanced technologies in cybersecurity while upholding ethical standards defined by law.

Policymakers should not only focus on individual nations as threats in the cyber world are global, hence they need to actively engage in international collaborations and forums aimed at establishing unified cyber policies and standards beyond defined borders. This also includes participating in diplomacy and treaty talks to enable a better regulated landscape that seeks to control the utilisation of AI in cyber warfare while encouraging collaborative efforts in cybersecurity.

At the same time, governments can take the lead in

defining the impact of AI technology in cyberspace to the public and educating the populace on the impact of AI technology for better precaution adoption.

In conclusion, sound management, strong policy formulation, and international collaboration can assist government officials and policy-makers in fostering a safe and morally aligned context for employing AI in cyber warfare, reducing possible dangers and simultaneously harnessing the powerful capacity of these technologies to improve cybersecurity on a national and global scale.

Corporate Responsibility and AI Integration

In the context of corporate-powered cyber warfare, businesses have substantial ethical and legal obligations regarding artificial intelligence (AI) systems integration and their applications. While businesses seek to improve their cybersecurity posture by adopting AI technologies, they must simultaneously observe ethical and legal boundaries governing the AI ecosystem. Corporate responsibility in AI integration goes beyond following applicable laws; it includes proactive engagement in ethical conduct, transparency, and responsibility in AI application for cyber defence. This chapter analyses various dimensions of corporate responsibility within the scope of AI integration in cyber warfare. Given that corporations hold sensitive information as well as critical national infrastructure, ethical use of AI is paramount to protect the integrity and security of digital ecosystems. Ethical considerations concerning the integration of AI touch on data privacy, algorithmic transparency, mitigation of bias, and the employment of AI for offensive cyber actions. In addition, businesses need to convene and collaborate with other relevant stakeholders, that include but are not limited to, government agencies, peers from the industry, and the general public to tackle issues posed by AI in cybersecurity.

The proper implementation of AI technologies requires a corporate culture that champions ethics and integrity while possessing a holistic comprehension of the implications of AI-powered cyber activities. Along with that, as organisations continue to adopt AI for cyber defence and threat intelligence, there is an increasing demand for proper control and governance of such AI systems. This means the formulation of adequate internal regulations, such as policies that outline the AI ethics principles, the governance structures for its implementation, and auditing frameworks for algorithmic assessments of the AI systems. In addition, proactive socio-technical measures of integrating AI necessitate funding towards employee education and perception campaigns on responsibly wielding AI tools, thus empowering employees with the requisite knowledge and skills on the responsible usage of AI technologies. Organisations can resolve stakeholder suspicion through strategically adopting corporate social responsibility tools, which help in the creation of a more trusted cyberspace. This is yet another reason stemming from the core argument of this chapter, that AI integration in businesses needs to be approached mindfully with the ethics of responsible innovation at the forefront. The same goes for the obligations to defend and digitally sustain firewalls around corporate assets.

Balancing Security and Privacy Concerns

The use of AI in cybersecurity has facilitated remarkable advancements in threat detection, incident response, and overall defence mechanisms. While these developments are certainly advantageous, they present difficulties in reconciling security needs with privacy issues. The deployment of AI technologies tailored to strengthen organisational digital perimeters creates additional risks to privacy that must be addressed. Therefore, while fortifying security infrastructures, intervening on privacy rights and freedoms should be avoided.

Data collection and its subsequent application frame the core of this challenge. Security systems powered by artificial intelligence require immense datasets to train and recognise patterns. However, gathering and processing this data raises pertinent privacy issues. Striking a balance entails implementing strong data governance frameworks that align with the principles of minimisation, purpose limitation, and transparency. Organisations must define and communicate the specific boundaries concerning the purposes of data collection, ensuring its utilisation is restricted to legitimate security concerns.

In conjunction with AI applications in cybersecurity, addressing security and privacy challenges requires the implementation of privacy enhancing technologies (PETs). With PETs such as differential privacy and homomorphic encryption, organisations can improve their security posture while protecting sensitive personal information. Through these technologies, companies can ensure that individuals whose information is utilised for security analytics remain anonymous, thereby protecting their privacy while providing invaluable insights.

Regulatory compliance, and compliance with the law on data privacy, marks a critical aspect of security and privacy balance in AI technologies used in cyber warfare. Companies deal with multiple legislative requirements like GDPR in the European Union and CCPA in California, USA. These laws come with strict data protection obligations which include limiting data retention, empowering data subjects, and employing privacy by design methods. Compliance with these laws highlights the need to provide security and privacy at the same time, further emphasising the legal interdependence of security and privacy.

From a broader perspective, the intertwining of artificial intelligence and cybersecurity presents an opportunity to consider the core values of an organisation, particularly requiring a shift to a holistic ethos that integrates proactive measures around privacy and security. Building trust and strengthening defences against malicious cyber threats is best achieved through a balanced approach to privacy and security. Adopting a privacy-by-design ethos anchored on ethical data management allows organisations the agility needed to withstand the onslaught of AI-driven cyber warfare.

Addressing Bias and Fairness in AI Systems

The equity and fairness of AI systems raises pertinent issues in cybersecurity. AI has the capacity to improve security, but it also comes with a danger of bias and discrimination if unmonitored. Bias can infiltrate AI at the data collection phase, algorithm design, and even at model training. Therefore, organisations must take steps to mitigate bias in AI to protect ethical standards and public confidence.

An important step in mitigating bias embedded in AI systems involves examining the data that trains such systems. Data-driven decisions based on flawed information can produce adverse consequences by aggravating discrimination and the sociocultural chasms of bias. Prevention and mitigation of any biases in datasets is crucial and needs to be addressed in profound detail by cybersecurity experts in data preparation workflows. Further, data provenance should receive equal attention as documents detailing the processes to promote accuracy and reliability need to be scrutinised as well.

The construction of algorithms and models needs to be focused on, as well as the information just mentioned, because they contribute significantly to the mitigation of bias. Decision logics and processes should be devoid of discriminatory practices, and algorithms designed need to reflect that. Fairness constraints and bias testing are examples of algorithms designed to measure and correct bias to ensure fairness.

AI systems demand incessant scrutiny not just on algorithms but also concerning bias detection and elimination throughout the system's life span. Evaluating AI-generated data for consistent validation can identify discrepancies and provide a means to resolve them, thus preserving equity in intervention. Feedback loops, while paired with regular audits, ensure the sustained integrity of objectivity and equity in cybersecurity AI systems.

An additional important factor when considering bias in AI systems is the inclusivity and scope of different perspectives in the design and use of these technologies. Policies to create inclusively structured teams with members from distinct disciplines are more likely to neutralise biased outcomes because they are able to consider and address myriad perspectives. Interdisciplinary input can strengthen assessments of AI systems and improve their fairness and inclusivity.

Ultimately, AI bias and fairness issues are central to ethical discourse involving AI trust, especially in the context of omnipresent AI in cybersecurity. Addressing bias at all levels of AI system development and operation makes these systems useful for security experts to exploit the advantages of AI with equity, no bias, and ethically correct mechanisms.

Public-Private Partnerships in Cyber Defence

In the context of cyber warfare, cooperation between the government and private organisations has become increasingly important. The nature of digital threats is fundamentally collaborative, and so is the need for government agencies and private sector organisations to cooperate in risk mitigation, threat intelligence, and the resilience of the nation's critical infrastructure and data assets. Fighting ever more sophisticated cyber adversaries who seek to exploit public and private sector weaknesses invokes the critical need for public-private partnerships (PPPs).

The ability to combine different sectors of knowledge and expertise in PPPs can lead to innovation and increase knowledge sharing, which sharply diverges from how government and public organisations work individually. Private companies bring with them experience and tech know-how, while governmental bodies offer intelligence, enforcement capabilities, and regulatory insight. Enhanced detection, response, and recovery from cyber attacks through AI is within reach with the combined efforts and resources offered by both sectors under PPPs, allowing the development and deployment of advanced cyber defence systems.

Moreover, PPPs enable advanced proactive participation through permitting the joint development of best practices, standards, and protocols which are prerequisites for the creation of effective cyber defence strategies. This joint participation permits the alignment of compliance and regulatory frameworks in cybersecurity within execution frameworks that are legislative and ethical while addressing the continuously evolving threat. Moreover, through joint cybersecurity exercises and training sessions, PPPs can prepare the public and private sectors to better address emerging cyber threats.

The timely reciprocal sharing of threat information and intelligence between public and private entities is vital for mitigating cyber risks and neutralising them before they escalate into attacks. The prompt exchange of emerging attack vectors, malware signatures, and anomalous behaviours within a network improves integrated multi-dimensional defence systems by allowing preemptive action to be taken on sophisticated systems and networks. PPPs also facilitate the response and recovery of incidents in a co-operative manner, thus enabling faster response time and reducing the effects of cyber incidents on businesses, critical infrastructure, and society.

The development of PPPs in the context of improving defence strategies and dealing with evolving cyber threats is crucial in today's world. PPPs deal with trust, transparency, and communication, which creates a resilient ecosystem that can regulate and tackle emerging cyber threats. In the end, both the public and private sectors working in cyber defence demonstrate a commitment to the protection of digital domains and cyberspace for coming generations.

Ethical Challenges of Offensive AI Capabilities

AI technologies intended to go on the offence are emerging with new evolving capabilities and offensive potentials, posing profound ethical considerations that need to be examined. The potential application of AI in offensive cyber operations raises issues surrounding the potential for AI to autonomously make decisions concerning acts of cyber warfare. Primary issues under ethical offence considerations are issues relating to absence of human supervision involved in actions offensively undertaken by AI technologies. This has an impact on the performance and the AI's operations that are conducted without any human input bearing serious consequences on collective humanity.

The use of AI technology in cyber weapons raises concerns regarding equity and the possibility of needless destruction. There is an urgent need to set boundaries with respect to the use of AI algorithms designed to engage in offensive operations while respecting the laws and sociopolitical frameworks related to conflicts between nations. The ethical issues of violence and the risk of collateral damage must be addressed.

Another ethical challenge is the concern of offensive technologies being AI-enabled and the subversion of such systems. AI algorithm weaknesses, coupled with adversarial vulnerabilities, pose significant ethical problems for trust and the reliability of AI systems to perform offensive tasks. Addressing these problems does require focus on policy issues, technological innovations, and ethical evaluation to curtail the use of manipulation in warfare through AI systems.

Furthermore, the social repercussions of offensive AI technologies also invite multidisciplinary ethical scrutiny concerning culpability, trust, and the dynamics of trust in progressively more fractured digital environments. The intertwining of AI with cyber offensive strategies serves the objectives of an adversary, highlighting the need for more comprehensive discussions from multiple perspectives about the moral essence of embedded ethics in order to formulate suitable policies and practices.

This, in turn, contributes to the creation of a strategic multi-tier approach directed towards the integrated offensive AI and cyberspace capabilities that is fully informed from the perspectives of law, ethics, and engineering, aiming for more responsible governance regimes that protect human dignity and rights, avoid flagrant abuse, and preserve social responsibility amidst defensive policies aligned with the use of AI and cyberspace hybrid offensive capabilities. Addressing these multidisciplinary perspectives in a preemptive manner contributes to building a favourable autonomous environment towards the guiding principles of ethical governance, transparency, and compliance in prospective AI-infused cyber warfare scenarios.

Future Directions in Legal and Ethical Standards

Considerations relating to legal and ethical challenges in the future of technology hinge upon the governing frameworks of cyber warfare due to the bolt-like growth of AI technology. As we develop the next generation of legal principles, it will be important to perceive different elements that are expected to impact the application of AI capabilities in cyber warfare.

The fluidity of so many factors moving in a particular direction indicates that the complexity of constellations in legal matters will change from this point onwards, dominating the reality around us. We begin from the ceaselessly developing advancements of AI. Self-demonstrating active AI systems encompass distributed control. Tempestuous AI driving cyber outcomes need different approaches in prevention and restriction based on ethical principles. Those that conflict with the laws of an open democratic community blending politics and law require international consensus. Another pivotal aspect, alongside ethics, touches upon the criteria of accountability as well as the criteria of involvement, to which artificial intelligence cyber technology will pay special attention in the future. A clear identification of responsibility and acceptance of transparent obligations under all aspects and stages of the AI cyber weapon's development, implementation, and use are necessary preventive measures and actions aimed at removing any unforeseen ethical dilemmas. This calls for the active creation of an autonomous model that indicates responsibility and obligation matrices for states, private companies, as well as for the creators of the AIs in question.

Additionally, while addressing the cyber aspects of AI, there is an increasing need for specific laws that cover the issues of data storage, consent of data owners, and its protection. Finding the balance between protecting the rights of private individuals while granting autonomy of action to AI to deal with potential cyber threats remains a serious ethical and political question. It will require consideration of several 'out there' ideas to remedy both these requirements – the means to achieve them are ethical transformations of the laws governing cyber warfare with the use of artificial intelligence technology.

When considering the future of laws and ethics, one must ponder the influence of AI in cyber warfare. The consequences for cyber employment, human rights, and international stability require attention. This includes considering the ethical concerns of humans being replaced by AI in automated defensive strategies and the offensive AI mechanisms concerning diplomacy and international relations, international politics, and conflict management.

Finally, concerning AI-driven cyber warfare, the legal and ethical frameworks of concern should focus on devising systems that allow continuous monitoring and change. These systems will regularly monitor the effectiveness and ethical concerns associated with the existing laws and identify where changes are needed to enhance them. With the adoption of a static legal culture, AI technologies and future challenges will be complied with; thus making the legal and ethical frameworks more resilient as they are continuously revised.

11
The Future Landscape: Towards Autonomous Cyber Conflict

Introducing Autonomous Cyber Conflict

The autonomous systems shift the balance of power within the context of cyber conflict because they can operate and function independently of human input. It is uncontested that autonomous systems will shape the future of warfare with their set capabilities. The concepts supporting autonomous cyber conflict stem from the benefits provided by quickened operational pace, enhanced speed of decision-making, and instantaneous processing of considerable amounts of information. The capability to decisively transform the character of cyber warfare catalyses the main development feature of these systems. Having unprecedented capabilities augments offensive and defensive cyber operations while adapting to multifaceted and constantly shifting threats. Autonomous systems promise proactive and adaptive responses to sudden cyber threats, greatly fortifying critical infrastructure and national security. It is essential to understand the strategic shifts that will arise from these systems and acknowledge how they shape conflicts sparked by emerging technology.

Furthermore, grasping the ethical, legal, and operational ramifications of autonomous cyber capabilities within the context of their use is crucial to navigating strategy. Essentially, analysing the concepts of autonomous systems and their autonomous relevance in potential future cyber conflicts presages their comprehensive assessment.

The Evolution of Cyber Warfare Technologies

The evolution of cyber warfare technologies has been relentless and rapid over the past few decades, driven by strong increases in computing power, networking capabilities, and digital systems on the global stage. Cyber warfare tactics used to depend on simple denial-of-service and malware attacks. However, due to more advanced technologies like artificial intelligence (AI) and machine learning, the transformations cyber conflicts undergo now have become extensive.

One impactful tier of those modifications AI has been used for is in offensive and defensive cyber operations. AI tools have allowed adversaries to perform precision strikes on multi-layered networks tailored to every single target, breach critical components, and bypass obstacles to advance within systems at breakneck speed. On the defensive side, AI tools have helped security analysts counter cyber intrusions by enabling them to detect anomalies, identify and respond to threats in real-time.

Additionally, the growth of the Internet of Things (IoT) devices has increased the vulnerability of cyber warfare, enabling IoT systems and smart devices as well as critical equipment to be hacked. The way this system is set up blurs the conventional borders of warfare. As a result, there is potential for physically rooted cyber attacks to disrupt critical infrastructure services and trigger enormous disaster.

The evolution of cyber warfare technologies has also given rise to capabilities for asymmetric warfare. State-of-the-art technology and cyber warfare capabilities can now be accessed by even non-state actors. Cyber conflicts have been aided by the dissemination of information and social manipulations which paves the way for data-based warfare. This results in altering the public's view and creating havoc and conflict around the world.

At the same time, cyber espionage and sabotage facilitated by a nation-state continues to pose acute threats to public safety, able to endanger entire societies, leak classified intelligence, and destroy trust in vital democracy pillars. All these factors highlight the alarming importance of formulating effective strategies of cyber defence concerning modern technological advancements and government policies.

Understanding the history and timeline of technological advancements in cyber warfare gives us insight into the current landscape. From analysing past cyber warfare technology, we can anticipate the future of self-governing cyber disputes as well as the critical rules for protecting our digital information in an age of continuous change and disruption.

AI-Driven Strategies for Future Conflicts

In contemplating future AI-enabled conflicts, one cannot fail to note how profoundly warfare is evolving. The application of AI in the realm of military strategy and tactics presents both a tremendous opportunity and a challenge to military modernisation led by AI strategies. AI applications in military contexts range from autonomous systems to heuristic forecasting of enemy movements. AI augments warfighting by increasing the speed and precision of information acquisition, as well as processing it to enable real-time data-driven decision-making at every command level.

The automation of warfare brought about by AI enhances the development of flexible, adaptable, and reconfigurable strategies aimed at countering rapidly occurring threats and dynamic and unpredictable changes in the environment of conflict. Moreover, AI technologies facilitate warfare automation, such as designing simulations of multiple scenarios for composite courses of action evaluation, which provide a greater understanding of decision-making problems for strategists and war planners. AI adds value to human analysis by processing large datasets, which aids in tracing critical changes in conflicts and even identifying underlying causal factors in complicated social and technological systems of geopolitics. On the other hand, the military application of AI technologies raises crucial moral and legal challenges.

Delegating important responsibilities to autonomous AI systems provokes critical issues of accountability and the risk of unintended outcomes. Furthermore, aligning AI-powered planning and operations with international humanitarian law and other ethical standards is essential to avert unlawful use of force and excessive military action. As countries compete to advance and incorporate AI into military capabilities, there is an urgent need to defend these systems against malicious exploitation and cyber-attacks. The growing dependence on networked AI systems introduces new and elaborate weaknesses that hostile actors might exploit, causing strategic disruption or sabotage. There is an imperative need to secure AI frameworks and build robust AI infrastructures alongside proactive defence strategies against emerging cyber threats to mitigate these risks. Furthermore, active cooperation in formulating frameworks for international collaboration on the consequences of AI-led strategies is crucial. Addressing concerns over escalating tensions and fostering communication between states is necessary to prevent conflicts driven by misinterpretations or miscalculations informed by AI situational assessments. The ongoing evolution of AI as a key factor in future conflicts amplifies the necessity for comprehensive approaches that combine complex technological efforts with ethical policies and international regulations towards a more secure world.

Challenges in Predicting AI Behaviour in Warfare

The application of artificial intelligence (AI) in warfare is particularly problematic in forecasting the behaviours of conflict AI systems. One of the central issues within this field is the decision-making procedures concerning AI systems, which are relatively intricate and volatile. The deep learning algorithms that process and make decisions for AI systems are very different from human-facing decisions since they possess the ability to process data on a large scale and at speeds far beyond human comprehension. This situation raises the question of whether AI could be effectively trusted or relied upon in any rigorous task calling for predictability, and during warfare operations, profound AI operations would require a high level of certainty. The problems could be further exacerbated by AI's ability to change autonomously and unpredictably, which increases the difficulty in attempting to forecast what their reactions would be to the evolving terrain on the battlefront. The shifting landscape and conditions also pose challenges for warfare AI systems that need to be real-time learners. Freeze zones of AI with a human warfighter add a myriad of other complications in forecasting the combined functions of artificially and anatomically manned forces in combat. Predicting the overall system behaviour hinges on the interplay of AI and human elements.

AI decision-making systems are challenging enough to interpret and analyse in the context of warfare behaviour prediction. Different models, algorithms, and training datasets all contribute to varied outputs from AI systems. Further, there are clashes of morality complicating the responsibility of programmers, system operators, and military personnel for ensuring adherence to operational protocols of AI systems. Adherence to rules of engagement creates difficulties in predicting responses from AI to morally delicate or morally ambiguous scenarios. Lastly, attempts to forecast AI behaviour are undermined by the possibility that opponents will seek to exploit biases in AI systems. Attempts to manipulate the systems will either eliminate or compromise attempts to forecast behaviour with certainty. As with many facets of technology, attempting to predict the behaviour of AI systems in warfare will fall behind due to uncharted risks and threats emerging with an evolving capability of anticipating probable dangers.

Human Oversight and Control in Automated Systems

The matter of human oversight and control in the realm of automated systems becomes especially salient around autonomous cyber conflict. As artificial intelligence and machine learning continue to advance, there will be a greater need for human oversight and control systems. This part will discuss the issues and the most important factors involved in ensuring and maintaining human control over the technologies designed for warfare and AI. Oversight includes the responsibility of monitoring and managing the actions of autonomous systems to remain within ethical, legal, and strategic boundaries. It forms governance structures, policymaking systems, and oversight mechanisms put in place to ensure that undesired outcomes and the unintended consequences of the deployment of AI technology do not occur. Additionally, human control refers to the willingness and ability to assess, change, or turn off autonomous systems based on situational analysis, ethics, or other unpredictable situations that might arise. While alluring and efficient, autonomous cyber conflict may pose ethical problems and critical risks that require strong human control and vigilant regulatory oversight. As irreplaceable as AI and machine learning allied systems can be, human intuition is key in the sphere of cyber-human conflicts.

Furthermore, designing and implementing fail-safe mechanisms, kill switches, and other emergency protocols feature prominently within the sphere of human control in AI-enabled cyber warfare. These controls are critical to avoiding catastrophic failures, unlawful capabilities provision, or conflicts escalating beyond control. Moreover, encouraging a culture of responsibility, openness, and sustained collaboration between humans and machines reinforces autonomous systems as aids, rather than primary decision-makers in cyber defence and offence. The persistent evolution of cyber threats and the agile nature of AI technologies emphasise the need for human resilience and involvement within the context of automated cyber conflict. Lastly, the debate about human supervision and control in systems automation for the purposes of cyber warfare highlights the pressing necessity to find the balance between emerging opportunities brought about by AI technology and human control, ethical requirements, and moral prioritisation in executing cyber operations.

Potential Scenarios of Autonomous Engagements

The increasing opportunities for autonomous participation in cyber conflicts, especially in terms of the prospects it carves into strategic policies for nations, raise a multitude of professional considerations. One such scenario entails the use of autonomous AI systems for the defence of critical infrastructure against large, coordinated, sophisticated cyber attacks. Such AI systems would be capable of dealing with intricate threats in parallel, overriding attempts made to compromise critical infrastructure services through automation, thus ensuring minimal disruption to services such as power grids, financial systems, and communication networks.

In the same manner, autonomous participation could also be employed in offensive cyber operations where the capabilities of AI are used to identify adversary networks and disrupt them with great agility and accuracy. Warfare patterns change spectacularly because of this transformation, as human judgement traditionally used to direct operations is now supplemented with decision-making processes by intelligent systems working at unparalleled speeds, granting Axis powers access to entirely novel forms of competitive advantage and unprecedented dangers.

Potential concerns stem from the use of modern military drones and cyber-physical systems that incorporate AI technologies. These systems have the capability to autonomously locate and interact with designated targets, perform reconnaissance operations, and tactically adapt to adversarial challenges. Although the integration of AI technologies into unmanned systems could enhance efficiency and further reduce human risk in active operations, there are already critical ethical and legal implications arising from a lack of accountability and negative outcomes.

Furthermore, the introduction of self-directed engagements raises the risk of AI technology instigating cyber warfare independently, using a given set of goals or by adapting to changing circumstances. This level of autonomy allows AI systems to interact with one another in a complex web of multi-layered actions and reactions, resulting in heightened speed and unpredictability in cyber warfare. This also amplifies the need to understand the risks associated with autonomous self-directed engagements, including but not limited to loss of control, unforeseen escalation, and the risks of fully relinquishing control over AI systems.

To sum up, the scenarios that can result from potential autonomous AI activities during cyber conflict represent a new transformational change in warfare that creates new opportunities alongside severely troubling challenges. From the ethical, legal, and strategic dimensions, autonomous systems would need to be addressed. As these scenarios emerge, it is clear that there must be a joint consideration from policy and military planners along with technologists to deal with the effective and responsible use of autonomously intelligent systems within the scope of cyber conflict.

Integration of Conventional Military with AI Systems

The integration of conventional military with AI systems marks a fundamental shift in warfare. With battling technologies, there is a need for the integration of AI into operations of traditional armed forces. This shift is not only disruptive when it comes to tactics and command selection; it creates specific requirements that are complex and require careful design and execution.

The essence of human-machine collaboration combines traditional military practice with AI technology, reflecting the concept of human-machine teaming. Given the rapid evolution of AI technologies, autonomous systems are expected to supplement human operators—working alongside them rather than fully substituting their roles. Through aiding data evaluation, recognising concealed patterns, and offering intelligent decision-making support, AI empowers military personnel to achieve unparalleled levels of situational awareness and actionable intelligence within real-time frameworks.

Moreover, predictive analytics and foretelling threats are among the key advantages that armed forces gain through the combination of AI technologies with conventional military assets. In turn, this will provide operational effectiveness and improve the overall efficiency of resources being used. Ultimately, this aids in the reduction of collateral damage, as well as civilian casualties during military operations in sensitive regions. Furthermore, the military platforms integrated with AI technologies provide adaptive and dynamic responses to evolving threats, thus adopting a proactive approach to addressing complex multi-dimensional security challenges.

Combining the traditional military paradigm with AI technology poses unique ethical, legal, and operational challenges. Policy considerations about LAWS - The Ethical Implications of Lethal Autonomous Weapons Systems Legislative Gaps and Policy Issues Loopholes in

Governance LAWS – LAWS The Evolving Threshold of Warfare and Control Loss of Human Control: Engineering Society From Ethical Considerations to Policy Action – draw on frameworks which lack legislative guidance as to what borders LAWS cross and how they are to be governed. International humanitarian law continually faces challenges juxtaposed to advancing AI technologies, raising questions around accountability, responsibility, and compliance (or lack thereof). This exploration supersedes LAWS as it touches upon AI and military operations as a whole, transcending borders and legally binding rules of engagement.

Cybersecurity and operational resilience safeguard plausibility frameworks for the military's AI structure and their integration. Ensuring unbreachable AI systems amidst cyber warfare incursions protects sensitive data alongside the risk of its affected manipulations. Preserving integrity and fending off exploitative hostile opposition reinforces the necessity of robust interoperability through strict security protocols to circumvent obstruction.

To build interdisciplinary trust, security, and mission integrity among hierarchical divisions, robust comprehensive AI-centred training programmes acquainted with exposed personnel systems are crucial for amenable collaboration. Concurrently, the necessity of human supervision in systems with decision-making AI components cements ethical waters which uphold moral and legal foundations amid warfare.

To sum up, the application of AI technologies to traditional military systems marks a new turning point for the strategy and tactics used in modern warfare. Adopting and accepting this change is likely to require a more balanced integration of technological advancements and consideration of ethical dilemmas, legal boundaries, and operational frameworks. If properly integrated through thoughtful design, democratic governance, and principled regulation, the constructive collaboration of human and artificial reasoning stands to reshape the evolution of war more broadly.

Risk Mitigation in the Age of AI Autonomy

In the pivotal era of AI autonomy, the significance of achieving a seamless amalgamation of AI with military and defence systems comes with distinct possibilities in equal measure with unexplored challenges. As we step into a world where AI is relied upon for making critical decisions, particularly in the context of cyber warfare, the matter of risk mitigation demands immediate attention. The unpredictability associated with AI technology brings with it the risk of negative ramifications, including the prospect of fully autonomous systems unilaterally deciding the system-triggered conflicts or escalations without human oversight. To counter this, strong risk mitigation measures must be enforced to curtail the aforementioned worries. One important pillar of risk mitigation needs AI systems, which are to be employed in military and cyber defence domains created, to be put through rigorous, their testing and verification processes. These must include the testing of their autonomous AI systems for response to varying conditions scenario based tests to identify weak spots in the exploitable decisional algorithms, which need to be reprogrammed to reduce the chances of wrong decisions being made and enhance optimal decision-making processes. Moreover, these autonomous AI systems need to define policies that are algorithmic and ethical to avert undesired losses due to malfunctions.

Collaborative work at international levels to define ethical guidelines and protocols dealing with the application of AI technologies in military operations can greatly reduce risks. Furthermore, increasing the diversity of AI technologies and integrating fail-safes and fallback options can provide ways to minimise damages from unforeseen AI actions. In addition to the mentioned technological measures, focusing on the human element in AI autonomy is equally or more important for risk reduction. Designing specialised training and forming multidisciplinary teams with AI, cybersecurity, ethics, and international relations specialists can foster an all-encompassing approach to risk evaluation and management concerning AI autonomy. Using transparency and explainability regarding AI processes for decision-making is crucial to foster proper understanding and trust by stakeholders which, in turn, increases risk awareness and mitigation strategies. All in all, the era of AI autonomy presents challenges that require an all-encompassing approach—technological, ethical, and strategic—to effectively address risks stemming from the use of AI in cyber warfare and military activities.

Collaborative Defence: International Cooperation Against AI Threats

The ever-escalating issue of AI-driven threats to cybersecurity calls for international collaboration, which can only be best achieved through cooperative defence. As AI autonomous systems advance, security paradigms may find it increasingly difficult to keep pace with the magnitude and complexity of AI-facilitated attacks. In resolving the issues of cross-border cyber warfare, a global coalition of states, institutions, and security professionals is essential.

Cooperation on an international level entails certain critical elements. First of all, it determines the policies and structures that need to be set up to provide timely exchanges of information on threats, such as communication and exchange frameworks. Preemptive action to curtail emerging AI threats can be taken proactively if advanced warning systems are improved through timely exchange of AI threat intelligence.

Moreover, collaborative defence involves creating responsive strategies and policies for the ethical and responsible use of AI technologies in cyberspace. International treaties are pivotal in establishing norms and AI deployment regulations, which will help in the governance of technologies while preventing adversarial outcomes.

The establishment of a worldwide consortium of cybersecurity specialists and scientists could stimulate AI-driven cyber defence collaboration on a global scale. This includes initiatives such as collaborative research, the establishment of digital repositories and platforms for sharing expertise, as well as advanced virtual training and teaching systems aimed at adding value to AI threat detection, analysis, and mitigation.

Shared or collective defence concepts are not limited to one country and encompass publicly funded private technology companies, partnerships with universities, and government branches. Stakeholders are bound to come up with more sophisticated cyber defence systems powered by AI when working collaboratively as they would incorporate different perspectives and resources, thus enabling them to beat more sophisticated attacks.

Furthermore, international collaboration is crucial to address the legal and policymaking silos pertaining to AI use in warfare and cybersecurity issues. Uniform legal policies and approaches towards the use of technologies across borders would enhance the governance of AI, encourage openness, and reduce legal ambiguity associated with collaborative defence operations.

In conclusion, adopting collaborative frameworks to counter AI-driven threats transforms the global cyber defence paradigm from reactive to proactive. This transformation allows a united global front to attract AI-hosted assaults on critical national infrastructure, sensitive information, and digital assets more effectively shield them from autonomous robotic warfare repercussions.

Conclusion: Preparing for the Autonomous Era

While flowing with the tide of an autonomous cyber conflict, a need to prepare and change becomes vital. Nation states and international bodies working together is crucial in strengthening our shields against impending doom caused by AI. Even though collaborative efforts are underway, it is apparent that the inability to foresee sooner the implications of the forthcoming self-governing era poses harrowing challenges. Here, I summarise the main conclusions I have drawn from the previous discussions, which serve as the basis for framing the subsequent autonomous future. First and foremost, developing international agreements and implementing strong laws and protocols which govern the behaviour set down on using AI in the cyber scope is imperative. Agreements must focus on granting peace and trust while also enabling a bailout from looming AI-driven attacks. In parallel, there is also a profound need for continuous conversations among the leaders of the governments, business sectors, and universities to keep pace with the swift shifts and changes in the autonomous cyber world.

To protect cybersecurity on a global level, organisations need to refine their policies based on cutting-edge technology. Furthermore, the swift adoption of AI in numerous industries, including defence, necessitates urgent investments in workforce education and talent development tailored to managing AI systems. Inclusivity can enhance collective intellect, inspiring greater innovation and adaptability. Ethics are equally vital in designing autonomous cyber technologies. Ethical guidelines must govern AI systems and regular independent reviews should safeguard against unintended consequences stemming from autonomous functions. A culture of cybersecurity and trust is critical as we prepare for the fallout from advanced AI on cyber conflict. This requires ensuring transparency regarding systems and processes in use. Safeguards must be put in place to maintain trust. Openness toward sharing information, auditing algorithmic bias, and checking for errors are essential for AI-enabled defences to remain credible and effective.

Proactive preparation along with cross-sector collaboration, ethical leadership, and transparency are precursors to readiness in standing at the cusp of the autonomous era. They together enable harnessing the profound impact of AI and easing its risks while paving the path toward a safer and more resilient future in the domain of cyber conflict. Embracing these principles will propel us forward.

Appendices

Glossary of Key Terms

The introduction laid out the basic concepts regarding the role of AI in digital infrastructure security. Familiarising oneself with the array of terms and ideas that constitute the spine of the rapidly evolving domain of artificial intelligence and its applications in cybersecurity is pivotal when venturing into this field. The goal of the current glossary is to aim towards providing a compilation of all relevant terms starting with basic definitions, giving an overview of crucial machine learning, neural networks, anomaly detection, adversarial attacks and more. All users of the glossary are presented with confidence and clarity towards the cyber defence space by focusing and capturing the cyber nuances. A true edge AI offers automated solutions efficient for user per AI CLI, but staggering for PDE or Python code explanations model. Stated simply, the glossary caters to precision and accessibility, serving distinct primal spark conversations on developments rich in neglected data that justify to the user they are knowledgeable about the driven AI cybersecurity. Further, through ease, AI is demystified building a universal language while compacting critical discussions in the cybersecurity space for nodes of knowledge through and around the world.

With diverse sources and industry standards, this glossary utilises a single, disciplined system of thinking alongside a synthesis of multiple fields, exemplifying the comprehensive nature of AI-integrated security practices. Understanding key terms fosters deeper comprehension of the complexities of cybersecurity, executive policies, and the evolving technologies of AI-based cyber defence systems. In other words, AI and cybersecurity professionals, and even enthusiasts, will together advance their collective knowledge through this glossary, which goes beyond standard definitions by actively enriching the dialogue on the subject.

List of Acronyms and Abbreviations

API: Application Programming Interface
CDN: Content Delivery Network
DNS: Domain Name System
DDoS: Distributed Denial of Service
EoP: Exploit of Privilege

HTML: HyperText Markup Language
IoT: Internet of Things
LDAP: Lightweight Directory Access Protocol
MITM: Man-in-the-Middle
OSINT: Open-Source Intelligence
PKI: Public Key Infrastructure
RAT: Remote Access Trojan
RSA: Rivest-Shamir-Adleman
SaaS: Software as a Service
SQL: Structured Query Language
SSH: Secure Shell
TLS: Transport Layer Security
UEFI: Unified Extensible Firmware Interface
URL: Uniform Resource Locator
VPN: Virtual Private Network
XSS: Cross-Site Scripting
Zero Day: Vulnerability That is Unknown to the Vendor

Technical Specifications and Standards

Compliance with prescribed requirements in the area of AI in cyber defence is of paramount importance as these ensure functionality in system integration, uniformity, and quality. In this part, we examine the prerequisites, critical technical prerequisites, and standards of cyber security powered by artificial intelligence systems.

As an example, we may mention the global ones, like ISO/IEC 27001 which is concerned with "Information Security Management Systems" and offers a governing structure for an organisation to efficiently manage and safeguard its information assets. Complying with such standards ensures that the provided AI-based cybersecurity solutions undergo rigorous controlled processes and adhere to best practice benchmarks.

Moreover, we can mention other authors like the NIST which significantly contributed to the set of guidelines and standards for AI systems by publishing the Special Publication 800 series. These documents examine domains such as cybersecurity, privacy and risk management, and provide a guide on how to integrate AI into cybersecurity solutions while maintaining ethical and legal frameworks.

Alongside these standards, regulatory bodies and sector-specific consortia often detail prerequisites concerning the use of artificial intelligence in cyber defence. ENISA, for example, provides in-depth technical directions relevant to the employment of AI within the EU cybersecurity framework, citing compliance with the principles of transparency, accountability, and the robustness of AI-powered security measures as mandatory.

In addition, the Association contributes to the development of deep specifications concerning the systems of AI and machine learning. They also address other concerns such as privacy in data, fairness of algorithms, and the often-overlooked AI cyber ethics which are critical in the creation of responsible and efficient AI systems.

Both AI and cybersecurity are evolving fields, thus, the technical specifications and standards need to be evaluated and modified constantly. This also entails having a comprehensive understanding of new technologies and putting new standards into place that would keep up with progress while being effective. Researchers, industry representatives, and government officials should work in tandem to be certain that the governing documents which have been put in place stand the test of modern, evolving cyber threats, both in relevance and functionality.

Following a set of standards, no matter how loosely determined, will work towards building the trust placed in confidence in AI-powered cyber security solutions. Following the standards set by the industry, organisations would be able to manage risk, lower costs, and enhance redundancy of systems while strengthening the security of the digital environment.

Resource Guide

As the landscape of AI in cyber defense and offensive strategies continues to evolve, it becomes increasingly important for professionals in the field to have access to a comprehensive resource guide that encompasses the latest tools, technologies, best practices, and expert insights. This section aims to provide a curated list of resources that cover a wide array of topics related to AI, cybersecurity, and digital warfare.

1. Industry-Leading Publications
Journals and Magazines
1. IEEE Security & Privacy Magazine
   - Focuses on AI in cybersecurity, privacy, and ethical considerations.
   - [IEEE Security & Privacy](https://www.computer.org/csdl/magazine/sp)

2. Journal of Cybersecurity (Oxford University Press)
   - Covers AI-driven cyber defense, threat intelligence, and policy.
   - [Journal of Cybersecurity](https://academic.oup.com/cybersecurity)

3. USENIX Login Magazine
   - Features articles on AI tools, machine learning algorithms, and cybersecurity practices.
   - [USENIX Login](https://www.usenix.org/publications/login)

4. ACM Transactions on Privacy and Security
   - Explores AI applications in security, including threat detection and response.
   - [ACM TOPS](https://dl.acm.org/journal/tops)

5. MIT Technology Review (Cybersecurity Section)
- Provides insights into emerging AI technologies and their impact on cybersecurity.
- [MIT Technology Review](https://www.technologyreview.com/topic/security/)

2. Research Papers
Machine Learning Algorithms in Cybersecurity
1. "Deep Learning for Cyber Security"
- Authors: Yishi Zhang, et al.
- Published in *IEEE Access* (2021).
- [Link](https://ieeexplore.ieee.org/document/9080294)
- Covers applications of deep learning in intrusion detection and malware analysis.

2. "Adversarial Machine Learning in Cybersecurity"
- Authors: Nicholas Carlini, David Wagner.
- Published in *IEEE European Symposium on Security and Privacy* (2017).
- [Link](https://arxiv.org/abs/1708.06131)
- Explores vulnerabilities in ML models used for cyber defense.

3. "Machine Learning for Detecting IoT Attacks"
- Authors: Hadeer ElKassaby, et al.
- Published in *IEEE Internet of Things Journal* (2020).
- [Link](https://ieeexplore.ieee.org/document/9053498)
- Focuses on ML-based threat detection in IoT envi-

ronments.

Threat Intelligence and AI
4. **"AI-Driven Threat Intelligence: A Survey"**
   - Authors: Mohamed Amine Ferrag, et al.
   - Published in *IEEE Communications Surveys & Tutorials* (2021).
   - [Link](https://ieeexplore.ieee.org/document/9349854)
   - Provides a comprehensive overview of AI in threat intelligence.

5. "Real-Time Threat Detection Using AI"
   - Authors: Arun Vishwanath, et al.
   - Published in *Journal of Cybersecurity* (2019).
   - [Link](https://academic.oup.com/cybersecurity/article/4/1/tyz004/5632201)
   - Discusses AI models for real-time threat detection and response.

Ethical Considerations in AI-Driven Cyber Defense
6. **"Ethical AI in Cybersecurity: Challenges and Opportunities"**
   - Authors: Luciano Floridi, et al.
   - Published in *Minds and Machines* (2020).
   - [Link](https://link.springer.com/article/10.1007/s11023-020-09539-5)
   - Explores ethical frameworks for AI in cybersecurity.

7. "Bias and Fairness in AI-Based Security Systems"
 - Authors: Inioluwa Deborah Raji, et al.
 - Published in *FAT* Conference (2020).
 - [Link](https://dl.acm.org/doi/10.1145/3351095.3372860)
 - Addresses biases in AI models used for cyber defense.

3. Reports and Whitepapers
1. Gartner Reports on AI in Cybersecurity
 - Provides insights into trends, tools, and best practices.
 - [Gartner Cybersecurity](https://www.gartner.com/en/information-technology/insights/cybersecurity)

2. McAfee Labs Threats Report
 - Highlights AI-driven threats and defense strategies.
 - [McAfee Labs](https://www.mcafee.com/enterprise/en-us/assets/reports/rp-quarterly-threats-mar-2023.pdf)

3. IBM Security X-Force Threat Intelligence Index
 - Analyzes AI's role in emerging cyber threats.
 - [IBM X-Force](https://www.ibm.com/reports/security-x-force)

4. ENISA Threat Landscape Report
 - Covers AI-driven threats and mitigation strategies.
 - [ENISA Reports](https://www.enisa.europa.eu/publications)

4. Expert Insights and Books

1. "Artificial Intelligence in Cyber Security: A Comprehensive Guide"
   - Author: James Scott.
   - Covers AI tools, techniques, and ethical considerations.
   - [Amazon Link](https://www.amazon.com/Artificial-Intelligence-Cyber-Security-Comprehensive/dp/1789950388)

2. "The AI-Powered Threat Intelligence Handbook"
   - Author: Dr. Mike Lloyd.
   - Focuses on leveraging AI for threat intelligence.
   - [Amazon Link](https://www.amazon.com/AI-Powered-Threat-Intelligence-Handbook-Machine/dp/1789137741)

3. "Weapons of Math Destruction"
   - Author: Cathy O'Neil.
   - Explores ethical implications of AI in security and other fields.
   - [Amazon Link](https://www.amazon.com/Weapons-Math-Destruction-Increases-Inequality/dp/0399590505)

5. Organizations and Think Tanks

1. Cybersecurity and Infrastructure Security Agency (CISA)
   - Publishes reports on AI in cybersecurity.
   - [CISA Resources](https://www.cisa.gov)

2. SANS Institute
   - Offers research papers and whitepapers on AI-driven cyber defense.
   - [SANS Reading Room](https://www.sans.org/reading-room/)

3. MITRE Corporation
   - Known for the MITRE ATT&CK framework and AI research.
   - [MITRE Cybersecurity](https://www.mitre.org/capabilities/cybersecurity)

4. Cambridge Cyber Intelligence Centre (CUIC)
   - Focuses on AI and threat intelligence research.
   - [CUIC Research](https://www.cambridgecyber.org/)

This list provides a solid foundation for exploring the intersection of AI and cybersecurity. For the latest updates, consider subscribing to newsletters from these organisations and following key researchers on platforms like [arXiv](https://arxiv.org/) and [Google Scholar](https://scholar.google.com/).

www.ingramcontent.com/pod-product-compliance
Lightning Source LLC
Chambersburg PA
CBHW051537020426
42333CB00016B/1973